Reed Carnrick

An abstract of the symptoms, with the latest dietetic and

medicinal treatment of various diseased conditions

Reed Carnrick

An abstract of the symptoms, with the latest dietetic and medicinal treatment of various diseased conditions

ISBN/EAN: 9783337201142

Printed in Europe, USA, Canada, Australia, Japan

Cover: Foto ©berggeist007 / pixelio.de

More available books at **www.hansebooks.com**

AN ABSTRACT OF THE SYMPTOMS,

WITH THE LATEST DIETETIC AND MEDICINAL
TREATMENT OF VARIOUS DISEASED
CONDITIONS.

THE FOOD PRODUCTS.

DIGESTION AND ASSIMILATION.

THE NEW AND VALUABLE PREPARATIONS
MANUFACTURED
BY
REED AND CARNRICK.

NEW YORK:

REED AND CARNRICK, 447 & 449 GREENWICH STREET.

1891

The Knickerbocker Press, New York
G. P. Putnam's Sons

PREFACE.

The marked favor with which our *Pancrobilin* has been received by the medical profession, as shown by the large number of letters received, citing the good effects that have followed the administration of this preparation, and the large and steadily-increasing demand from the trade, has caused this further advance by us in practical and scientific pharmacy.

Manufacturing chemists have previously placed their preparations in the hands of the Profession, simply stating their general characters, and in a common way recommended their use in almost all diseases. This has left the busy practitioner to find out for himself when they could be used to advantage and when best to omit their administration.

In this book, we have not only announced our new *Pancrobilin compounds* and *Tablet preparations*, but have given for each its exact physiological actions, its therapeutic advantages, and stated definitely when, why, and how it should be given to obtain the best results.

In the dietetic treatment of disease, when the ordinary food-stuffs cannot be taken or digested, we have recommended our prepared and pre-digested food products, because we are confident that they are not equaled by any other food product in the market. All of our products are highly nutritious, easily digested and assimilated, and correspond more closely with the present chemical and physiological standards than those of other manufacturers. Consequently, their use will enable the system to obtain the largest amount of nutriment with the smallest expenditure of vital force.

Recognizing the great importance of a thorough knowledge of the comparative values of the ordinary food-stuffs, of the physiology of digestion, assimilation, and oxidization, and of the functions of the bile and pancreatic fluid, a special section, Part II., has been written exclusively upon these subjects.

This section is an answer to the many inquiries received from all parts of the country for fuller information in regard to these all-important conditions.

Part II. has been prepared with great care and somewhat in detail, and we have quoted freely from the instructive course of lectures*—delivered by Prof. William Henry Porter at the New York Post-Graduate Medical School and Hospital during the winter of 1890 and 1891—because in these lectures the functions of the liver, bile, and pancreatic fluid are fully and concisely stated.

Professor Porter, in his paper on "Fel Bovis Inspissatum,"† has also shown how necessary it is to remove the old and deteriorated bile from the system before introducing a new supply, and we in turn have emphasized this fact both in Part I. and Part II.

In Part II., the composition and comparative utility of all kinds of the common food-stuffs are clearly outlined, both from the physiological and from the practical or clinical standpoint. In this section the dangers arising from false feeding are illustrated, the methods by which diseases are developed clearly stated, and the complete functions of the bile and pancreatic fluid given.

Under "Treatment" the necessity and the best method for removing the old bile is described. The *Cholagogue Tablet* and the *Pancrobilin* or Bili-pancreatine compounds are fully explained.

We have prepared several *Pancrobilin compounds*, also a number of valuable *Tablet preparations*, giving to each a specific name by which it can be differentiated from all other preparations in the market. This has been done so that the physician can be positively sure of obtaining those manufactured by us.

We further desire to state that these *Tablets* are prepared from the very best materials that the market affords

---

* Published in *The Medical News*, Philadelphia, January 10 and 24, and February 28, 1891.

† Published in *The Medical News*, Philadelphia, May 2, 1891.

and will always be found uniform in strength and to contain exactly what the formula states should be contained in them.

They are prepared in this *Tablet* form instead of a pill, as tablets are now considered the most efficient form for administering this class of medicinal agents.

Our *Pancrobilin preparations* are prepared with the greatest of care so that the physician can always be sure that he has a thoroughly reliable article. For it is absolutely necessary that the ox-bile and pancreatic extract remain unchanged and free from all extraneous matter, otherwise failure must follow where success ought to crown the efforts of the physician.

Great care has to be exercised in the selection, preservation, and preparation of the inspissated ox-bile, otherwise many, if not all, of its essential qualities will be completely lost.

When failure has followed this line of treatment, careful investigation has invariably shown that the preparation of bile and pancreatic extract used was defective in quality, which was the true cause of the unsatisfactory results obtained.

We have found that, by exercising proper care in the selection of the ox-bile and pancreatic extract used, they can be made into pills and coated with gelatine and still retain all their virtues for months, and so far as known ox-bile will keep in this condition for years without undergoing any appreciable change and without losing any of its physiological activity ; while the ordinary inspissated ox-bile of commerce quickly deteriorates unless it is rapidly used or great care is exercised in its preservation after the original package has once been opened and the contents exposed to the atmosphere.

Cognizant of the great importance of bile as a therapeutic agent, and fully aware of the difficulty experienced in securing a reliable inspissated ox-bile even in our large cities, we have undertaken the manufacture of these *Pancrobilin* or Bili-pancreatine compounds, so that the profession can always have at its command a thoroughly reliable preparation and one that does not easily deteriorate.

Without desiring to trench upon the field of the physician, or to depart from the highest standards as manufacturing chemists, we feel confident that our new departure is a decidedly scientific advance, and one that will be welcomed by the medical profession as timely and valuable in the treatment of disease, and that it will ultimately result in great good to the human race at large.

Correspondence from physicians on the subject-matter of this book will be cordially welcomed by us and will receive prompt attention.

Samples of our new *Pancrobilin compounds* and *Tablet preparations* sent on request.

REED & CARNRICK,
MANUFACTURING CHEMISTS,

OCTOBER, 1891. New York City.

INTRODUCTION.

To the Medical Profession.

Gentlemen :

The object of this little pamphlet is to fill a long-felt gap in medical literature, and to place in the hands of the active practitioner, in condensed form, the more recent views in relation to digestion and assimilation, and the most valuable dietetic and medicinal treatment in use for disease.

It has been divided into two parts, for greater brevity and usefulness.

In Part I. a concise outline of the leading symptoms of each disease is given, also the dietetic and medicinal treatment that is most effectual in relieving or curing these abnormal conditions.

The subject-matter has been so arranged that the ever-busy practitioner can at once turn to whatever disease he may be interested in, and within a very few seconds have at his command the most useful dietetic and medicinal line of treatment at present known to us.

In Part II. a more accurate but concise account of food is given, together with the methods by which many of the diseased conditions previously enumerated are developed. Also the complete proof substantiating the truth of the many unqualified assertions made necessary by the brevity practiced in the former section is here clearly stated.

Respectfully,

REED & CARNRICK

October, 1891.

CONTENTS.

PART I.

In this section the more recent methods for successfully dieting and treating the following conditions are briefly outlined.

PART II.

PART I.

DYSPEPSIA AND GASTRIC DERANGE-MENTS.

Under this heading of gastric dyspepsia may be included all the acute inflammatory conditions of the stomach.

The most common symptoms pointing to a purely gastric affection are pain at the time of eating, belching up of alkaline or acid gases or fluids, nausea and vomiting ; all occurring at the time of attempting to eat the meal or immediately after the ingestion of the food.

This form of difficulty is best treated by abstaining from all kinds of solid food and the use *per os* of cracked ice, Kumysgen, Carnrick's food, lacto-cereal food, or lacto-preparata.

As the stomach becomes stronger, peptonized milk, then ordinary milk, eggs, and lean meat should replace the fluid preparations, and finally the vegetable compounds may be admitted.

As most of these acute gastric affections are developed after prolonged over-feeding and the previous establishment of intestinal and hepatic derangements,—as fully explained in Part II.—a cholagogue tablet at night and the mild Bili-pancreatine pill or the liquid Bili-pancreatine should be given before each meal, so soon as the acute gastric irritability has been subdued. This line of treatment will in many instances prevent a return of the attack and establish good health.

Cocaine in one-grain doses every three or four hours will often quickly relieve the nausea, vomiting, and pain.

INTESTINAL INDIGESTION OR DYSPEPSIA.

From the well-established fact that the bulk of the digestive work is done beyond the pylorus, or in the intestines, disturbances which take their origin in defective work at this point and within the entero-hepatic circulation are among the most common and troublesome that come to the general practitioner for relief or advice. In fact the great mass of all our diseased conditions take their primary origin at this point,—and the methods by which these abnormal conditions are developed are fully explained in Part II. Suffice it to say, that when the biliary and pancreatic secretions have for any reason become impaired or defective in quality or quantity, the work to be accomplished at this point is incompletely performed. Abnormal fermentative changes are induced, irritating gases are formed, the pabulum to be absorbed poor in quality, and the patient goes steadily from bad to worse. In some instances the symptoms are mild, but in other instances they become so aggravated that the suspicion of the existence of a malignant disease is often entertained.

The chief symptoms indicative of these entero-hepatic imperfections are a heavily-coated tongue, the tendency to bloat with gas from one to four hours after taking the meal. With this, there is much abdominal distress, at times reaching a stage of acute pain in the pit of the stomach and leading to the erroneous diagnosis of a gastric affection. Finally, after a varying period of distress, a reverse peristalsis is established, the accumulated gas is forced back into the stomach and a variable quantity is belched up, which is soon followed by relief from pain and abdominal discomfort. If the bile produced is defective in quality and quantity, there is constipation and clay-colored stools ; if the quantity is large but the quality defective, there is usually frequent burning and tarry stools.

When the hepatic functions are largely at fault, the skin will assume a dirty, thick, and muddy hue, and the urine will be scanty, strongly acid, high-colored, and contain an abnormally small amount of urea and an unusually large quantity of uric acid or other by-products.

The most effectual way to treat these affections is to cut down the quantity and regulate the quality of the food taken. At first, for one or more weeks, the diet should be restricted to Kumysgen, Carnrick's food, lacto-cereal food, and lacto-preparata. Then, so soon as the composition of the urine changes for the better, eggs, lean meat, and fish may be allowed ; finally bread, and lastly, a limited supply of the vegetable food substances should replace the prepared and liquid food products.

When a perfectly satisfactory improvement has been established, these individuals should be cautioned against eating too freely of all kinds of food and particularly of the starchy and sugar-containing classes, otherwise a speedy recurrence of the unpleasant symptoms may be expected. A study of Table I. in Part II. will easily enable the selection of a suitable cereal or vegetable diet for the most perfect and safe kind of nutrition.

As the biliary and pancreatic fluids are more positively defective in this class of affections than in the gastric, more active medication is required to effect a cure. The old and deteriorated bile must be fully removed from the system before commencing the regular dietetic course. This can be effected in a variety of ways ; with a full dose of calomel, with twenty grains of blue mass and a compound cathartic pill administered together, with a full dose of podophylin, or by a saline, as phosphate of soda. But the best results are obtained by the use of two cholagogue tablets every night for three nights in the mild cases, or by two of these tablets administered every four hours until the bowels commence to move freely, in the severe instances.

While the cholagogue tablets are being taken, the patient must also be given one of the compound Bili-pancreatine pills three times each day. If three meals are being taken, the pill, or liquid Bili-pancreatine should be administered about fifteen minutes before eating ; when upon an exclusive fluid diet the remedy should be taken about the same time before taking a regular allowance of whatever fluid food may have been selected.

With a great deal of abdominal pain and an unusually large quantity of gas, the following pill will be found most serviceable :

"ANALGESINE TABLET."

℞

Hydrochlorate of Cocaine	20 grains
Phenacetine	60 grains
Exalgine	60 grains
Salicylic Acid	60 grains

M. Divide into forty tablets, of which from one to three should be taken every four hours until the pain is relieved.

In all these cases, there is more or less nervous depression, feeble heart action, and an abnormally low arterial tension. To overcome these conditions, to raise the arterial pressure, to improve the heart's action and augment the nervous innervation distributed to the digestive organs in particular, the following is specially adapted :

"INNERVATINE TABLET."

℞

Sulphate of Strychnine	1 grain
Hydrochlorate of Caffeine	60 grains
Extract of Damiana	90 grains
Extract of Taraxicum	30 grains

M. Divide into sixty tablets, of which from one to two should be taken every four or five hours.

GASTRIC ULCER.

The round perforating ulcer of the stomach is most fre quently seen in young and poorly-nourished females. The most characteristic symptoms are sharp, lancinating pain, referred to the epigastrium, shooting through to the back and down into the abdomen. The pain is greatly increased by the ingestion of food and by pressure. Vomiting is almost always present and pronounced, and is aggravated by the introduction of anything into the stomach. Vomiting of bright red or clotted blood is well marked in one-fourth the cases. There are also well-defined symptoms of gastric and intestinal indigestion.

At first the stomach must be given absolute rest. This is best accomplished by the use of nutritive enemata and the abstinence from all kinds of food, either liquid or solid, by the mouth. From two to six ounces of Kumysgen, Carnrick's food, lacto-cereal food, or lacto-preparata, peptonized milk, meat broth peptonized, eggs peptonized, oysters peptonized, etc., should be injected into a perfectly cleansed rectum every four or six hours.

For internal medication, one-half to three-grain doses of cocaine, nitrate of silver or iodoform, either alone or in combination, administered in pill form three or four times daily, will afford the most relief. Hydrocyanic acid dilute, or opium may be needed to relieve the pain.

As soon as the stomach will retain anything, small and frequently repeated quantities of Kumysgen, Carnrick's food, lacto-cereal food, or lacto-preparata should be given and finally regular food, as before indicated.

When food is taken by the mouth, the mild Bili-pancreatine preparation, either in pill or liquid form, must be given, to improve the digestive and nutritive functions.

GASTRIC CARCINOMA.

Cancer of the stomach usually occurs between the age of forty and sixty and most frequently in men. At first the symptoms are those of chronic indigestion. The more positive indications are a poor appetite, nausea, and vomiting of large quantities, especially the coffee-ground-like liquid, pain of a dull, gnawing character, a well-defined tumor, pyloric obstruction, and the cachectic appearance.

With all these signs well defined and a lack of improvement under Bili-pancreatine pills, the diagnosis is almost certain. In cases of doubt, and in those instances of pyloric obstruction with rapid emaciation and no sign of tumor, modern science indicates an early exploratory laparotomy, followed by pylorectomy or gastro-enterostomy, as the case may demand.

In most instances in the way of treatment, surgical interference is the only line of action that offers any satisfactory relief, and this must be undertaken early to secure the most benefit.

The diet should be Kumysgen, Carnrick's food, lacto-cereal food, or lacto-preparata, with Bili-pancreatine in pill or liquid form, to establish the highest digestive and nutritive condition possible with the least irritation.

HEPATIC DERANGEMENTS.

Under this heading, the symptoms are about the same as have already been given in connection with the aggravated forms of entero-hepatic indigestion on page 8.

a. With simple functional derangement of the liver, however, there is a stronger tendency to become jaundiced, but without complete jaundice. If there is a defective production and discharge, both in the quantity and quality of bile, there is marked constipation, pronounced intestinal indigestion, rapid emaciation, and the discharge of urine containing more or less bile pigment and salts.

Where the quality of bile discharged is defective and the quantity abundant, the movements of the bowels are too frequent and irritating. There is marked evidence of intestinal indigestion, flatulence, etc. These patients, however, do not emaciate, but their adipose and muscular tissues become very soft and flabby. They become physically and mentally weak, and great sufferers from insomnia, mental delusions, and hypochondriasis. Their urine is scanty, strongly acid, high colored, loaded with uric acid, urates, or phosphate (usually both), bile pigment and salts.

b. With the marked jaundice, there is simply a more aggravated stage of one or the other of the above conditions.

c. When biliary calculi are formed they are usually developed secondarily to the above mild state of functional derangement, and often do not make their presence known until they are found at the autopsy. In other instances they become dislodged, and in attempting to pass through the gall duct to the intestines give rise to intense colicky pains, profound nausea, vomiting, and jaundice. If found in the stools, the diagnosis becomes absolutely certain.

The production of these conditions of functional derangement of the liver is fully explained in Part II.

The dietetic treatment consists in a short course of Kumysgen, Carnrick's food, lacto-cereal food, or lacto-preparata, either alone or with a limited amount of solid food, until the functional activity of the digestive organs is fully re-established.

The medicinal treatment consists in the persistent use of the cholagogue tablets and the Bili-pancreatine pills or solution. The old and deteriorated bile must be completely swept out of the system and a new supply constantly furnished until the system is perfectly capable of manufacturing the proper quantity and quality of biliary and pancreatic fluid to perfectly accomplish the digestive and assimilative work of the animal economy. The cholagogue effect will in many instances have to be repeated several times to fully reach the difficulty. Attention to diet and the use of the plain, strong, mild, or compound and tonic Bili-pancreatine pills will have to be continued for months to effect a permanent cure.

For the immediate relief of a biliary colic and the removal of the stone, chloroform in half-drachm doses has been injected into the gall-bladder through the abdominal wall. Failing in this, laparotomy is in order. After the abdomen has been opened, a number of procedures have been recommended, all of which, with the exception of two, when frankly stated, have as many failures as cures to their credit.

The two operative methods which offer the best returns are laparotomy and the injection of chloroform through a fine needle into the duct just behind the stone (this will usually cause the stone to soften and rapidly disappear), or the duodenum may be opened by a linear incision opposite to the side of entrance of the common bile duct, and then, the bile duct dilated, the stone extracted, whole or after being crushed, the fragments removed and the intestinal and abdominal wound closed — as successfully practiced by Prof. McBurney.

The re-formation of biliary stones can be prevented by properly dieting in a manner similar to that in use for the prevention of their formation.

d. The functional disturbance which occurs in connection with chronic diseases of the liver must be treated in a similar manner, but a little less actively, excepting in cirrhosis and syphilitic affections.

The cholagogue tablets being administered every second, third, or fourth night, and the Bili-pancreatine preparation as usual, t. i. d.

If the organic lesion be syphilitic in origin, an active cholagogue action, followed by the compound Bili-pancreatine pills, and the free use of the iodides will almost completely eradicate the defect in a short space of time.

In the old alcoholics with cirrhosis of the liver and defective biliary and pancreatic secretions, the free use of the cholagogue tablets and the daily administration of the simple or compound Bili pancreatine pills, or the fluid preparation, will remove the jaundice and regulate the deranged digestion almost like magic.

If, on the other hand, the neoplastic growth be sarcomatous, carcinomatous, adenomatous, or lymphomatous in nature, neither medicine nor the surgeon can effect a cure, but the patient must be made as comfortable as possible until death gives the only and permanent relief. The nutrition can be largely sustained, however, and the patient made comfortable by the use of Kumysgen, Carnrick's food, lacto-preparata, or lacto-cereal food, and the Bili-pancreatine pills.

THE URIC ACID CONDITION.

This frequent abnormality is simply a form of intestinal and hepatic indigestion which is brought about largely by eating too freely of all kinds of food, but particularly of the vegetable classes, which, by containing a large percentage of proteid matter and a super-abundance of starch, tend to continually over-tax the oxygenating capacity of the system, until finally the proteids are imperfectly transformed and uric acid in abundance appears in the urine.

This condition usually assumes one of three types :

a. The simple form, in which the chief symptoms are mild digestive disturbances, with the frequent and well-marked brick-dust deposits only.

b. The gouty form, in which the soda salts and the uric acid-forming elements are united in the various structures of the body and form the gouty concretions. These deposits are most frequently located in the cartilages covering the ends of the bones entering into the formation of the metatarso-phalangeal joint of the great toe. They may, however, be deposited in the structures of the shoulder-joint, the wall of the stomach, the kidneys, or in any tissue of the body. At the time these deposits are being precipitated, we have the characteristic gouty attacks of intense pain and localized inflammatory changes, with an absence of the uric acid and urates in the urine. These symptoms continue for a few days, when the local symptoms completely subside and the urine again contains an excessive quota of uric acid and urates.

c. The same uric acid state, but with the formation of calculi in the pelves of the kidneys, which is due to some condition of the pelves of the kidneys and their contents by which the microscopic crystals of uric acid coalesce into macroscopic masses.

The movement of the larger stones in the pelves of the kidneys causes the localized renal symptoms, while the obstructed descent of the smaller calculi through the ureter gives rise to excruciating pain along the course of this duct and down the inner side of the integumental surface of the thigh, with a marked contraction of the scrotum or labia on the afflicted side. Bloody urine is often discharged during the passage of the stone, especially if the transit be prolonged, and always after the final escape of the stone from the urinary passages. All of these are known as characteristic symptoms of renal stone or colic.

The dietetic treatment in all these conditions consists in putting the patient upon a limited and fluid diet, for a time at least, preferably Kumysgen, Carnrick's food, lacto-cereal food, or lacto-preparata, then milk, eggs, meat, bread, and finally vegetables, taking care at all times to give no more than can be perfectly digested and assimilated. This can be ascertained with precision by a close study of the urine.

The medication consists in the free use of the cholagogue tablets and the Bili-pancreatine pills or fluid preparation. The nervous system can be strengthened by the use of the "innervatine" tablet recommended on page 10.

The formation of uric acid and renal calculi are almost always prevented by the internal administration *per os* of one grain doses of carbolic acid three or four times daily. The presence of this alcoholic compound in the system appears to change the oxidization processes in such a manner that a soluble by-product is formed instead of the insoluble uric acid, and these recurring attacks of renal colic are at once arrested, until a suitable diet with proper medication can be made to completely eradicate the defective condition from the system.

For the temporary relief of the acute attacks of gout nothing relieves so quickly and effectually as a free catharsis by the combined action of the cholagogue tablets and a saturated solution of magnesium sulphate, the former administered every three hours until three or four have been given, and then drachm doses of the latter every half hour until the desired effect is fully established. Following this, ten minim doses of tr. colchicum which has been made from the green plant, every three or four hours ; also the Bili-pancreatine pills three times a day before the meal.

THE OXALIC ACID CONDITION, OXALURIA.

This common condition is characterized by the oxalate of lime crystals being found in the urine in varying quantities, either alone or in combination with uric acid. It is a disturbance in anabolism, indicated by a degree lower in the oxidation series than in the uric acid, and oxalic acid is one more of the by-products formed by incomplete proteid metabolism. It is a well-established law that as the uric acid condition grows worse it becomes an oxaluric state, and as the oxaluric condition disappears, a uric acid stage intervenes before a complete cure is effected.

This particular form of imperfect oxidization appears to cause considerably more nervous irritation than the uric acid state, in consequence of which neurasthenic symptoms are more or less prominent and require attention.

Here, as in the former condition, a few days or weeks of an exclusive diet, composed of Kumysgen, Carnrick's food, lacto-cereal food, or lacto-preparata every two hours, followed by milk, eggs, meat, bread, and vegetables in small quantities, is the most effective diet.

The cholagogue tablets and Bili-pancreatine preparations as before stated; the compound and tonic pill with constipation, and the mild form, otherwise. The sulphur-tartrate tablets are also very serviceable in this condition.

Ten minims of dilute nitro-hydrochloric acid after meals is often of service.

As there is considerable lack of nervous energy and a hyperaesthetic condition of the system, the innervatine tablets given on page 10 will often be needed.

THE LACTIC ACID CONDITION, OR RHEUMATISM.

This form of abnormality is represented by a still lower degree of oxidization of the proteid bodies and one in which lactic acid is formed as one of the by-products developed along this line of nitrogenous metabolism. It is largely produced by an over-indulgence in the vegetable class of food-stuffs and repeated exposure to wet and cold. It may, however, be produced by over-indulgence in meat alone, but this more frequently results in the gouty condition.

As a result of this disturbance of the perfect anabolic processes of the body, some form of katabolin is formed out of the proteid substances, and this is the antecedent to lactic acid found in the excretions. It is this katabolin which, while circulating through the system, excites the local congestions or truly inflammatory changes in the fibrous structures of the body and gives origin to the acute and painful symptoms commonly known by the universal title of rheumatic affections.

The urine in these cases is also apt to be scanty in quantity, strongly acid in reaction, high-colored, with a deficient quantity of urea, and an increase in the uric acid urates and phosphates, and it may also contain some lactic acid.

It is the katabolin, however, formed out of the proteid substance and antecedent to the lactic acid found in the excretions, that is the intrinsic and damaging element, because lactic acid as such is a comparatively inert CHO substance and cannot exist under its own form in the blood.

To prevent the development of rheumatism and remove the tendency to have recurring attacks, a strict course of dieting must be instituted and continued for months. Such patients should first be placed on a liquid diet, such as Kumysgen, Carnrick's food, lacto-cereal food, or lacto-preparata, taken every two or three hours. After a few days or weeks (depending upon the indications found in the changed urine), milk and eggs, then meats of all kinds, finally bread and a limited amount of vegetable food-stuffs, may be added to the diet. Potatoes, however, should be scrupulously avoided, because clinical experience has repeatedly demonstrated that one full meal of potatoes will quickly disturb the normal proportions existing between the formation of urea and uric acid and, certainly within a few hours after the ingestion of the meal, precipitate an abundant deposit of uric acid in the urine, with a rapid development of all the vague and unpleasant symptoms which are common to this form of nutritive disturbance.

During the acute attack a free cholagogue action must be established with the cholagogue tablets and the strong or compound and tonic Bili-pancreatine pills. After this the salicylic acid or its sodium salt, pushed to the full limit, will reach most of the cases and quickly subdue the intensity of the attack. If the salicylic compound is not well borne by the stomach, ten minims of the oil of wintergreen every two hours is quite effectual. The analgesic pill already referred to on page 10 will often relieve the pain. When chronic, the sulpho-tartrate tablets, the alkalies, or the iodides, will yield the best results, — two to four tablets at night or night and morning. The alkalies and iodides may be given in from ten to sixty-grain doses t. i. d. They have to be given in large doses and persistently to produce marked effects. Dieting and attention to the intestinal and hepatic functions are absolutely essential. Without the bile pills the other remedies have comparatively little effect.

NEURASTHENIC CONDITION: NEURASTHENIA.

The definition given for this malady by Professor Dana* is, "a morbid condition of the nervous system, whose underlying characteristics are excessive irritability and weakness."

Neurasthenia appears to take in a host of indefinite and variable symptoms, in which all the functions of the body are more or less disturbed, but in which an enfeebled nerve innervation is always present. The entero-hepatic functions are almost if not always at fault ; assimilation and oxidation are defective, and the urine abnormal. This being true, the dietetic treatment plays an important part. These patients should have all the out-of-door exercise that their weak condition will permit without marked fatigue. The diet should commence with Kumysgen, Carnrick's food, lacto-cereal food, and lacto-preparata, then, gradually added, milk, eggs, meat, and fish of all kinds, and, finally, vegetables in moderation.

The cholagogue tablets can be used every second or third night, or once a week, as the depleted condition of the system will stand ; then tonics, such as the corrigent pills, one or two, t. i. d., with the pill for innervation on page 10. Massage is also of service in many instances.

The bromides can be given, but they simply produce temporary quiet, and as soon as the drug is discontinued the symptoms speedily recur and usually in an aggravated form, — so that this line of treatment is palliative, while the former is curative, in nature.

To produce sleep, the following tablets will be found the most serviceable :

"HYPNOTINE TABLET."

℞

Abstract of Hyoscyamus	3 grains
Chloralamide	60 grains
Sulphonal	60 grains
Extract of Damiana	90 grains

M. Divide into sixty tablets, of which from one to three may be taken every three or four hours.

---

* Dana: *The Post-Graduate Journal*, N. Y., January, 1891, p. 26.

THE DIABETIC CONDITION, OR DIABETES MELLITUS.

In some instances this malady appears to be quite clearly associated with some nerve lesion, but in the vast majority of cases it is positively traceable to over-indulgence in foods of all kinds, but particularly those of the vegetable class. The reason why most vegetables and cereals are apt to over-tax the digestive and oxygenating capacity is, that they contain such a large percentage of proteid matter and an excessive quantity of starch. The proteid of the vegetable being a higher nitrogenous compound than that found in the animal kingdom, it is harder to peptonize, and requires more oxygen for its perfect anabolism. The starches have to be transformed into glucose and then to alcoholic-like compounds before they can become available for use by the system, all of which requires a large outlay of vital energy and the expenditure of much oxygen. For a time Nature puts up with the imposition, but after awhile it yields to the strain. The starches and sugars continue, by the unchangeable laws of chemistry, to be completely transformed into carbon dioxide and water. But in the more difficult task of changing the proteids Nature falls short, the nitrogenous bodies are imperfectly transformed, and glucose-forming elements are developed out of the proteid substances, which, when they reach the renal organs, are finally converted, by the action of the epithelial cells lining the uriniferous tubules, into a complete glucose product, and discharged from the body as such with the urine.

This method of explaining the pathology of diabetes clearly illustrates the reason for the almost universal hepatic and renal lesions found at the necropsy made upon diabetic subjects.

This line of reasoning positively indicates the plan of treatment to be followed.

Cut down the quantity of food taken and regulate the quality. But as physiology and chemistry teach that, in health, a well-regulated mixed diet is the most valuable, so in diabetes mellitus the same law holds true.

Far better results can be attained by cutting down quantity and allowing a small allowance of the sugars and starches than by attempting to institute an exclusive proteid diet.

But for a short time they should be placed upon the simplest kind of diet, to give the system, as nearly as possible, a period of complete rest. This is best accomplished by taking Kumysgen, Carnrick's food, lacto-preparata, or lacto-cereal food in moderately small quantities. These food products, being highly nutritive and easily digested, will enable the system to appropriate a large amount of nutriment with a small expenditure of energy. This will cause a rapid fall in the density of the urine and in the glucose and uric acid, with an associated rise in the percentage of urea.

After a week or two the solid food-stuffs may again be added, such as milk, eggs, meat, and fish of all kinds, bread, macaroni, and a small allowance of the least starchy cereals and vegetables.

By this line of dieting and suitable medication, a diabetic can be made practically well and will live from five to forty years, but most of the time a little glucose will be found in the urine.

If the condition be detected early and properly dieted from the inception of the disease, a complete cure may be expected.

In the way of medication, the cholagogue tablet occasionally, with one of the Bili-pancreatine pills or the solution, may be given as indicated by the digestive and urinary symptoms — the mild pill if the bowels are moving daily, and the strong or compound and tonic when constipation is present.

The sulpho-tartrate tablets will also be found of value, and in cases of much nervous depression the innervating, pill; also the corrigent pill and cordial analeptine are highly recommended.

NEPHRITIC LESIONS, OR BRIGHT'S.

Lesions of the kidneys, commonly called Bright's disease, and characterized by albumen and casts in the urine, include a dozen or more quite distinct and separate conditions,* but the laws governing their production and treatment are largely the same.

Briefly stated, they are either due to local congestions, direct injuries, or the imposition of an unusually large amount of work upon the renal glands at a time when their nutrition is defective or not proportionally sustained. This is most frequently brought about by over-eating, imperfect digestion and assimilation, the introduction of some foreign body, as the infectious poisons which disturb the anabolic process of the body and increase the imperfection of the nutritive changes, and cause to be developed an abnormal amount of imperfect and effete material. This waste matter, together with the original poison, has to be cast out of the system largely by the renal organs. At the same time their nutrition is below normal and their work largely in excess of the usual requirements. As a result, the structural elements of the kidneys undergo simple metamorphosic changes, or truly inflammatory lesions are established, followed by abnormalities in the urine, chief among which are changes in density and the presence of albumen and casts. The proteid matter, instead of being converted into complete excrementitious products, like urea or uric acid, stops short in the form of a readily diffusible albuminoid.† In this way we find a rational explanation for the absence of solids and the presence of the albumen in the urine.

Sudden or prolonged exposure to cold will also produce renal congestion and excite structural changes in the renal organs in a similar manner.

Cognizant of the methods by which renal lesions are developed, the treatment consists in decreasing the amount of excretory work to be accomplished and improving the nutritive state of the kidneys.

---

* Porter : "Renal Diseases and Urinary Analysis." Wm. Wood & Co., N.Y., 1887.
† *Ibid : The Post-Graduate Journal,* N. Y., January, 1891, p. 13.

This is best accomplished by first placing the patient upon the Kumysgen, Carnrick's food, lacto-cereal food, or the lacto-preparata, thus raising the nutrition to the highest standard and decreasing the amount of work thrown upon the renal organs by restoring the disturbed anabolic process of the body nearly or up to the normal standard. As the digestion and assimilative powers improve, milk, eggs, meat, and fish may be added to the diet, also bread and vegetables in moderation.

In this class of lesions the cholagogue tablets and Bili-pancreatine preparations seem to have, in conjunction with the above diet, almost magic powers in staying the progress of, and curing absolutely, the renal condition.

All substances, both nutritive and medicinal, which can in any way irritate the excretory organs must be scrupulously avoided.

When the urine is scanty or likely to be suppressed, the following pill is specially applicable, being prompt in action and non-irritating :

"DIURETICINE TABLET."

℞

Sulphate of Strychnine	1 grain
Hydrochlorate of Caffeine	60 grains
Powdered Digitalis	30 grains

M. Divide into 60 tablets, of which from one to three may be taken every three or four hours.

The corrigent pills are especially applicable as a tonic in this class of diseases.

CONSTIPATION.

Under this heading a whole chapter might easily be written, but as constipation is such a common complaint, and especially so among women, its symptoms need not be described.

Its successful treatment depends largely upon its cause, but from the well-known fact that a defective supply of bile is at the foundation of all these cases, the following pill more nearly fulfills the physiological indications than any that has been previously recommended :

"COMPOUND AND TONIC PANCROBILIN PILL."

℞

Inspissated Ox-gall	80 grains
Pancreatic Extract	20 grains
Sulphate of Quinine	40 grains
Extract of Nux Vomica	12 grains
Compound Extract of Colocynth	40 grains
Extract of Damiana	60 grains

M. Divide into forty capsules, of which one to three should be taken three times daily, before meals.

To successfully overcome this troublesome condition, three things are necessary : the use of the above pill, outdoor exercise, such as walking or horseback riding, and, above all things, the prompt attention to Nature's indication that she is ready to evacuate the contents of the lower bowel. The common recommendation that a set hour is the only time at which Nature should be allowed to relieve herself, is as radically wrong as to compel a person to micturate at a specified time and only at that hour. The whole study of the animal kingdom indicates clearly that both the bladder and lower bowel should be promptly emptied whenever Nature gives the signal. When this rule is rightly obeyed, constipation rarely if ever occurs, unless the entero-hepatic functions are decidedly at fault.

DIARRHŒA.

This condition, like constipation, would easily admit of a whole chapter being written upon the subject, and yet the affection is too common to require a detailed account of its symptoms.

Diarrhœa may be subdivided into as many kinds as there are causes. But, with all, an undue number of watery stools characterizes the affection. With this, there may or may not be pain.

The treatment consists in first removing all the substances contained in the alimentary canal and then, if possible, sterilizing the same. The latter can practically be accomplished by the use of sterilized drinks and food products only.

A properly-regulated diet is the first thing to be considered. To begin with, Kumysgen made with sterilized water in sterilized bottles is the surest way to secure a perfectly sterilized food product. The Carnrick's food, lacto-cereal food, or the lacto-preparata all mixed with sterilized water in sterilized cooking utensils will, as far as possible, shut out all contaminating and irritating substances. The alcohol generated in the development of the Kumysgen is specially beneficial in stimulating the depressed nervous system so commonly associated with these affections.

In children, the Carnrick's food will yield almost marvelous results where all other means have failed.

In some cases it may be necessary to limit the food to raw chopped or scraped beef, but in the majority of instances the above preparations will meet every indication.

For the immediate checking of the diarrhœal condition, a combination of opium, nitro-hydrochloric acid, and camphor as follows, will yield the most satisfactory results :

℞

Deodorized Tincture of Opium	2 drachms
Dilute Nitro-Muriatic Acid	6 drachms
Camphor Water	16 drachms

M. One-half to one drachm every two or three hours.

On account of the great nervous depression associated with this class of diseases, some form of camphor is absolutely essential.

With the actions of the bowels under control, the mild Bili-pancreatine pill or liquid preparation should be given three times daily to prevent the secondary constipation and enable Nature to most perfectly digest the food-stuffs used and thus prevent a recurrence of the attack or the degeneration of the acute into a chronic diarrhœal state.

APPENDICITIS, TYPHLITIS, AND PERITYPHLITIS

Under these three titles is included a variety of inflammatory changes implicating one or all the anatomical structures in the region of the right iliac fossæ. The morbid action will range in severity from a mild inflammation of the appendix vermiformis and the cæcum or tissue surrounding them up to that of suppuration and the formation of intra or extra-peritoneal abscesses and a gangrene of the appendix.

The symptoms associated with these lesions may start as a mild chill, a little elevation of temperature, some colicky pain in the region of the right iliac fossæ, with mild diarrhœa or constipation. These symptoms may last a few days and pass off, and every sign of disturbance subside, and perfect health be re-established.

In other instances the chill or chills may be more severe and persistent, the temperature rises high, with vomiting, great pain and tenderness, and a rapidly-developing fullness or well-defined tumor in the right iliac region. These symptoms may rapidly grow worse with clear signs of localized suppuration. Other cases run along for two or three weeks and then subside with recovery. There may be frequent recurrences in these signs until finally an acute and rapidly-progressing suppurative process is developed ; or there may be a slowly-forming tumor in this region without much constitutional disturbance until by the large size and the pressure effects of the neoplasm, nutrition is impaired. These large post-cæcal tumors often communicate with the lumen of the intestine and are a source of slow self-infection which steadily undermines the health.

In the most aggravated cases, there is a rapid gangrene of the appendix vermiformis with perforation into the peritoneal cavity, general infection of the sack, with collapse and death as the first and only signs and symptoms.

The treatment of these conditions has to be governed by the kind of condition present. If there is no marked sign of suppuration, the expectant plan with dieting and rest can be followed.

When suppurative signs declare themselves, an early operation under aseptic precautions is imperative. Unless the operation is done within three days in the acute class of cases the mortality is very high, but if done within those short limits the prognosis is favorable.

In the chronic cases with the formation of the post-cæcal abscesses the pathological cavity should at once be opened and free drainage established.

In all classes the diet should be fluid easily digested and that which will leave the least amount of refuse matter to pass through the lower bowel. For this purpose the Kumysgen, Carnrick's food, the lacto-cereal food, or lacto-preparata come the nearest to the physiological indications of any that can be found. Having overcome the local defect successfully, the patient should gradually be brought back to a well-regulated mixed regime.

The Bili-pancreatine pills or liquid will regulate the bowels, improve digestion and assimilation. When the nervous system is greatly depressed the innervating pills should be administered, together with the Bili-pancreatine pills.

COLITIS OR DYSENTERY.

Colitis includes a variety of pathological changes commonly called dysentery. The morbid processes occurring in the rectum or colon may be a simple catarrhal, follicular, or croupous inflammation, which may go on to the formations of deep ulcerations, perforations, or gangrene of the intestinal wall.

The condition may be acute or chronic, endemic, sporadic, or epidemic, but, in all, the characteristic and typical symptoms are the frequent and small bloody mucous squibs, with a great deal of pain, tenesmus, and tormina.

The endemic and sporadic cases usually develop as a simple diarrhœa, after which the typical local symptoms are developed, and finally those of a general constitutional character are established.

In the epidemic cases, which are supposed to be caused by the presence of a pathogenic bacteria, the constitutional symptoms come first as a severe chill, high fever, prostration, nausea, vomiting, the heavily-coated tongue, diarrhœa, rapidly developing into the typical dysenteric dejections.

In the management of these cases, the regulation of the diet is of first importance. As in diarrhœal affections, the diet should consist of sterilized food products, such as Kumysgen, Carnrick's food, lacto-cereal food, or lacto-preparata, all made with sterilized water and in sterilized cooking utensils. The water taken should also be sterilized. Sterilized Underwood Water comes as near to perfection as anything that can be found.

Here, as in all other abnormal conditions, the old and deteriorated bile must first be removed from the system and in its place a new supply temporarily furnished by the use of the Bili-pancreatine pills or fluid preparation. This will enable Nature, by her own methods, to keep the alimentary canal as nearly aseptic as possible.

For the direct improvement of the local condition, large doses of powdered ipecac will, in all instances, yield the best results. The ipecac should be given in doses ranging from forty to eighty grains. It should be administered in capsules, or, better still, in wafers, after which all fluids should be withheld for three or four hours. To prevent nausea or vomiting, a full dose of opium may be given about twenty minutes before administering the ipecac, and two hours earlier a small fly blister, placed over the scrobiculus cordis, will enable the stomach to retain the drug until it passes into the lower bowel. This large dose may be repeated on the second or third day. In many instances, two doses are sufficient. Of course, all the old-line remedies and the starch and laudanum injections may be administered. Flushing the rectum and colon with sterilized water or water containing some antiseptic will often yield good results, and very decidedly aid the ipecac treatment.

a. In this connection, that troublesome malady which used to be known as membranous enteritis, but which, more correctly speaking, is a membranous desquamation from the colon, deserves attention.

The characteristic symptom is the discharge with the stool, at varying intervals, of shreds, cords, or complete tubular casts of the colon. These bodies are formed in the colon as the result of a chronic congestion of its mucous coat and the production of an unusually large amount of thick, tenacious mucus, which forms a thin coating upon the inner surface of the colon, and finally is exfoliated in thin shreds or rolled up into cords or as a complete cast of the gut.

Associated with these periodic discharges, there is usually constipation, sometimes diarrhœa, abdominal discomfort or pain, functional disorders of the liver and small intestines, also symptoms referable to the uterus and ovaries; associated with these, we have evidences of hysteria, hypochondriasis, and neurasthenia.

These cases can generally be cured by a persistent course of dieting and medical treatment directed chiefly toward the digestive and hepatic derangements.

At first, for a limited period of time, the diet should be restricted to a small quantity of Kumysgen, Carnrick's food, lacto-cereal food, or lacto-preparata, taken at short intervals. Then, as the digestion becomes stronger and nutrition more effectually sustained, milk, eggs, lean meat, and fish of all kinds can be gradually added to the fluid food products ; lastly, bread and the more easily digested and least starchy vegetables.

Before commencing the dietetic course, the cholagogue tablets may be used for a day or two, also the Bilipancreatine pills, strong, mild, or compound, as the condition of the bowels may indicate. If there is much abdominal pain, with accumulation of gas, the analgesic pill (page 10) should be used. With considerable nervous depression the innervating pill (page 10) should be administered. Careful attention to the diet and nutritive processes, together with a persistent use of the Bili-pancreatine preparations will result in a cure in almost every instance. To accomplish the desired result, however, may take months of steady attention.

b. Carcinomatous disease of the colon is not infrequent, located either in the sigmoid flexure or at other points along the large intestine.

The characteristic symptoms are a localized pain and tumefaction at some point along the line of the colon, persistent constipation, followed by sudden attacks of diarrhœa ; the stools containing at these times blood, a puslike substance, and broken-down tissue.

Careful dieting will give great relief at times, but early surgical interference is the only procedure that offers any prospect of permanent relief.

FÆCAL IMPACTION.

This condition frequently occurs in old people. They complain of constipation or diarrhœa. They lose flesh and strength and die from fæcal poisoning when not relieved.

In adults the same thing may occur at some point above the rectum. These cases are usually constipated, and suffer from colicky pain, flatulence, and a tumor is easily detected. In rare instances they may have localized pain, tenderness, fever, and prostration.

In exceptional instances the patient has diarrhœa, looses flesh and strength, and dies. At the necropsy, the only abnormality is an over-distended colon packed full of fæcal matter.

The fel bovis alone, or perhaps better in combination, as in the following formula, will soften and remove quite effectually impacted fæces and stimulate the lower bowel to action when all other means have failed. It is also a very powerful agent for bringing away the gas and relieving troublesome tympanites :

℞

Inspissated Ox-bile	1 ounce
Glycerine	4 ounces
Castor Oil	2 ounces
Water	8 ounces

M. Sig.—This added to a pint, or, better still, a quart of warm soapsuds; the larger amount can be retained when slowly injected into the lower bowel.

After several copious injections have been administered and retained for a time and have come away without producing the desired effect, it is justifiable to assume that the colon is free. This has been verified with sufficient frequency in the dead-house to warrant the statement that it will invariably soften and remove fæcal matter from the colon. Excepting, of course, a tight stricture of the colon or an occlusion of the lumen of the gut, which cannot be passed by the enema.

PERITONITIS.

This includes all forms of acute and chronic, toxic and non-septic inflammations of the peritoneum. It may be a localized or diffuse process.

The symptoms will vary with the extent of peritoneum involved and with the cause of the inflammation. If it be a localized and simple process there will be a slight chill, a little rise in bodily temperature, and localized pain of moderate intensity at the seat of inflammation.

If the inflammation be general or septic in nature, there is usually a severe chill, rapid rise in temperature, great abdominal pain, and distention with gas. They vomit first the contents of the stomach, then a spinach-green substance, and finally the contents of the bowels. The pulse is rapid and wiry, but later on it becomes small and feeble. The countenance has an anxious expression, the respirations are shallow and imperfectly performed. The patient lies upon the back with the thighs flexed upon the abdomen and the legs upon the thighs, in this manner protecting the abdomen from all pressure or injury. The bowels are usually obstinately constipated except in the septic cases, where there is a tendency to diarrhœa. Any one or two or more of the typical symptoms may be absent and still the necropsy show the lesions characteristic of peritonitis.

The diet should be fluid in character, composed of Kumysgen, Carnrick's food, lacto-cereal food, or lacto-preparata. The quantity taken should be small and at frequent intervals.

Two plans of treatment are employed, that known as the Prof. Alonzo Clark treatment, which consists in dieting and bringing the patients fully under the influence of the powdered opium and keeping them there until the improvement was marked.

The modern treatment, which is unquestionably the best in the septic variety, is to omit the use of the opium and to keep the bowels moving freely with a saturated solution of magnesium sulphate. To this may be added the analgesic pill and the Bili-pancreatine preparations, if the best results are to be obtained.

The local source of infection must also be treated antiseptically. For instance, if due to an abortion and retained secundines, they must be completely removed and the uterine cavity rendered thoroughly aseptic. The antiseptic line of treatment holds true for any wound that may have given rise to the toxic element which excited the peritoneal inflammations.

INFECTIOUS DISEASES.

Under this common title of infectious diseases, there may be included, at least in a general way, all those conditions which are developed in the human economy as the result of the introduction of some form of poison into the system. The toxic agent may be a microbe, a ptomaine, leucomaine, a toxine, a kataboline, anaboline, or an unknown quantity. As a result, however, of the entrance of the foreign substance, whatever its nature may be, there is produced at first a profound disturbance in the entero-hepatic circuit, and particularly in the bile-producing function of the liver. These physiological digressions are made manifest by a chill, rise in bodily temperature, headache and backache, a heavily coated tongue, constipation, with a slight tendency to jaundice, scanty, high-colored, and super acid urine which contains an excess of uric acid, urates, and phosphates, and a deficient quantity of urea. Up to this point, the symptoms are nearly the same for all infectious diseases, and it is not until the protecting power of the liver has been overcome and the infection been permitted to gain access to the whole system that specific symptoms are developed by which one disease can be isolated from another.

In common, then, up to a certain point, the treatment is the same for all infectious diseases regardless of a specified name. The diet also should be the same.

A fluid easily digested and a highly nutritious diet taken in small quantities and at short intervals will maintain the nutritive powers at the highest standard possible under these unfavorable circumstances. For the accomplishment of these ends, Kumysgen, Carnrick's food, lacto-cereal food, and lacto-preparata in perfectly sterilized form are especially adapted. They can be used either alone or with the other food products, such as milk and beef tea or broths.

In the way of medication, the first and most important thing to be effected is the removal of the old, deteriorated and poisoned bile. This can be successfully done in a number of ways as previously described on page 9, but the cholagogue tablet, variously used, most effectually accomplishes the end in view. In expelling this impaired

biliary fluid, much of the original poison is also removed,
which tends to lessen the intensity and shorten the duration
of the specific disease in question. Having removed the
old bile, a new stock must be supplied to the system by
giving the Bili-pancreatine pills three or four times daily.
By thus improving the antiseptic condition of the alimentary canal, improving digestion and restoring the assimilative and excretory powers as nearly as possible to the
normal standard, Nature has in this newly administered
bile, the most powerful and only true antipyretic measure
which has any claim upon physiological laws for its true
method of action. In so far as the temperature is dependent
upon these digestive laws, it will be completely removed,
but so long as any abnormal disturbance of the physiological economy is kept in motion by the presence of any
foreign element within the system, a certain amount of
abnormal temperature must be present, the removal of
which by powerful agents masks the true state of the
system, imposes extra burdens upon an already overtaxed
economy, and, as has been repeatedly proven, prolongs these
affections, raises rather than lowers mortality, and, as Ehrlich,
the great German clinician, practically puts it, kills the patient.

To devote a special section to each infectious disease
would be wasting valuable time in constant repetitions.
The common law that governs one governs all, and when
this fact is thoroughly appreciated and acted upon great
good will flow out to humanity at large.

a. Typhoid Fever in particular is especially adapted to
this line of dieting and medication by Fel Bovis. Having
its chief lesion in the alimentary canal, it has been found
that with bile alone more satisfactory results have been
obtained than with any other known method of medication.
The use of the Bili-pancreatine preparation is a step further
in the direction of perfection than the use of the bile alone.
Its use in typhoid fever has not been as extensively tried
as the bile alone, but so far as used Bili-pancreatine has
been equally if not more satisfactory in its action.

Under the head of prevention the following rules must
be strictly enforced.

Have the fæces discharged into a vessel containing
several ounces of one of the following solutions. Thoroughly

mix with a still larger quantity before emptying into the common receptacle.

℞

Bichloride of Mercury	2 drachms
Hydrochloric Acid	10 drachms
Water	1 gallon
M.	

℞

Bichloride of Mercury	2 drachms
Permanganate of Potash	2 drachms
Water	1 gallon
M	

℞

Chloride of Lime	4 ounces
Water	1 gallon
M.	

The anal region should be thoroughly cleansed with the second solution after each dejection.

All soiled clothing should be soaked in one of these solutions, or be steamed and then thoroughly washed and boiled.

Great care should be exercised in the use of drinking water or milk in an infected locality.

b. Cholera Infantum. What has been said in discussing the infectious diseases in general applies here with equal force. The development of this affection appears to be predisposed to by warm, damp weather and a depressed condition of the nervous system. Its exciting cause is unquestionably due to the introduction of some foreign or irritating substance with the food taken, presumably bacterial in character, unquestionably a toxic agent so far as the normal physiological processes are concerned.

This poisonous substance produces both gastric and intestinal irritation, and cause nausea, vomiting, and diarrhoea, with rapid emaciation of the child.

These attacks vary greatly in severity and duration, ranging from a mild form of diarrhoea with slight gastric disturbances, lasting a few days or weeks and then terminating in complete recovery, up to a point in which all the symptoms may be intense, death ensuing in a few hours or days at the latest.

The treatment consists in removing the offending subtance, sterilizing the alimentary canal, nourishing the system and stimulating the depressed nervous system.

The foreign body which is the cause of the symptoms can, in a measure, be removed and the functional activity of the liver improved by a few and frequently repeated small doses of gray powder, calomel, or the bichloride of mercury, as the character of the symptoms may indicate, the latter being specially applicable where the stools contain much mucus.

Sterilization of the alimentary canal is absolutely essential. This is practically effected by giving only sterilized food products and injecting into the lower bowel quite large enemata of simply pure sterilized water or that medicated with some antiseptic such as the bichloride of mercury, boracic acid, etc.

The best form of diet is small and frequently repeated quantities of Kumysgen made perfectly sterile as before described, or Carnrick's food, lacto-cereal food, or lactopreparata. There are few cases that will not be more thoroughly improved by these preparations than by any other form of diet.

The Kumysgen is especially valuable because of the alcohol produced and contained in it. It is also very easily digested and a very highly nutritive food product, on account of the soluble proteids, and one which comes more nearly to perfection than any which has heretofore been furnished to the Profession.

As a rule the less medicine infants and children are given the better. The chief thing is the regulation of the diet.

If, however, the diarrhœa is very profuse and rapidly exhausting the child, it must be stopped. The following mixture will be found very available for this purpose :

℞

Camphorated Tincture of Opium	1 drachm
Tincture of Catechu	2½ drachms
Fluid Extract of Coto Bark	1½ drachms
Peppermint Water	2 ounces

Mix and give one teaspoonful every one or two hours.

In the most urgent cases, with constant nausea and vomiting, the sixth of a grain of gray powder on the tongue, and a warm starch water enema containing a little laudanum and acetate of lead, will often prove most effectual in arresting the discharges from the bowel.

A change of locality should always be recommended. If the attack occurs in the city or at the seashore, send the child at once to the inland mountains. If the attack occurs inland, the reverse should be the practice,—send the child to the seashore.

The gentle counter-irritation of a warm but mild mustard or spiced plaster over the belly should never be omitted. These should be changed frequently so that the abdomen will not be chilled but constantly kept warm as well as stimulated by their presence.

c. Tubercular Affections. Under this name we have a large class of lesions involving every organ and tissue in the body and causing the death of thousands almost daily. It is perhaps one of the most clearly established bacterial diseases that attacks the human race. At the same time there are two very important etiological factors which precede the action of the germ. They are a predisposition or mal-nourished state of the system and a local cause of inflammation, without which the germ part of tuberculosis makes but slow progress, if it can advance at all. This nutritive factor has very largely been lost sight of in the mad chase after the germ. Yet plain, old clinical facts constantly assert that if Nature's nutritive powers are properly adjusted, she will both keep out and expel the germs from almost any system to which, perchance, they may have happened to gain access.

If the bacillus tuberculosis is to be regarded as the sole cause of this class of lesions it is hardly possible to class tuberculosis as an inherited disease. For, if so, it must be assumed that the germ is transmitted from father or mother to the offspring and often lies dormant in the system of the descendant for years and then suddenly develops its deadly powers and rapidly destroys life.

On the other hand, false methods of feeding, and especially while the mother is carrying the child *in utero*, and the same line of false feeding from birth up to and through adult life, undermines the nutritive vitality and causes the system to become specially prone to the action of all forms of infection and especially so to that of the tubercular bacilli.

Observations among the animal kingdom show that the carnivora rarely have tuberculosis, while the vegetable feeders are particularly prone to this form of disease. Carnivora, while fed upon an animal diet, cannot be successfully innoculated with tubercular matter; confined and fed upon an opposite class of food products and the experiments rapidly become successful.

Tuberculosis in the human subject is most frequently found in starch and sugar-fed subjects and in those who are compelled to live in close quarters.

The only line of treatment that has yielded any satisfactory results has been the sending of such cases to the wild mountain regions where the diet of necessity is largely of the animal class, such as game and fish. With this out-of-door life, an almost exclusive use of an animal diet, and a moderate amount of alcohol and cod-liver oil to supply what CHO element the system absolutely demands, has resulted in the positive cure of a fair percentage of genuine tubercular subjects.

When in these cases the nutritive powers have reached a low ebb, and the stomach and intestines have become very irritable, Kumysgen, Carnrick's food, lacto-cereal food, and lacto-preparata will often furnish the system with a large nutritive supply when all other means have failed. They will furnish a nutritive nucleus out of which a final cure can ultimately be developed.

As a rule the less medicine given in these cases the greater will be the probability of ultimate recovery. The Bili-pancreatine pill should be given regularly to assist Nature to more perfectly digest the food products taken. The corrigent pills will act as a good tonic which is usually needed. A small amount of peptonized cod-liver oil and a moderate use of some form of alcoholic stimulation will usually be found of decided benefit. The lungs and air passages should be kept as free from the inflammatory products as possible, for in this way one large source of septic self infection and high temperature is abated.

d. La Grippe or Influenza. This infectious malady, which
has been so prevalent throughout the country during the
past two winters, is unquestionably due to the introduction
into the system of some form of toxic substance.

Epidemic influenza can be positively traced back to 1570,
since which time up to 1870 more than ninety well-defined out-
breaks have been recorded. The epidemics travel from
east to west, lasting about six or eight weeks in a given locality. In itself it is not generally fatal but increases the mortality of other diseases, thus raising the general death rate.
The attacks may be repeated in the same individual. The
disease is not contagious nor has any specific bacteria been
as yet demonstrated.

Its lesions are a catarrhal inflammation of the mucous
membrane, of the eyes, nose, ear, throat, larynx, bronchi,
stomach, or intestines. Its severity differs in each locality
and in the person attacked. In some, the bronchial effects
are the most severe, while in others the gastric symptoms are
the predominating manifestations. Thus different epidemics vary in intensity and in the part of the body involved. Lobar and broncho-pneumonia, pleurisy and em-
physema frequently complicate the disease.

The characteristic symptoms are a chill, rise in temperature, intense frontal headache lasting for days, intense
pains all over the body, and great bodily prostration ; — this
later symptom being out of all proportion to the local manifestations.

The tongue is coated, the bowels constipated, the urine
scanty, strongly acid, and high colored. In fact, all the symptoms common to the infectious disease in general are present.

The treatment consists in dieting with Kumysgen
and milk for a day or two. Free catharsis with the cholagogue tablets and the use of the strong or compound and
tonic Bili-pancreatine pills. For the pain, the analgesic pills,
and for the prostration, the innervating pills. For the local
treatment of the upper air passages frequent spraying with
benzoinol, medicated with cocaine, menthol, oil of eucalyp-
tol or cubebs, or salol, gives the largest amount of relief.

The complications must be met as they arise, but when
uncomplicated, the severity of the attack ought to be over
in from three to five days.

There is, however, great mental and physical weakness often lasting for weeks after the acute symptoms have entirely passed.

This is best overcome by a change of climate, the use of Kumysgen with a plain, nutritious diet, the mild or strong Bili-pancreatine pills three times a day before meals and the innervating or corrigent pills as may be indicated. With much anæmia the latter should be selected. In some instances, peptonized cod-liver oil and alcoholic stimulation will be required.

CYSTITIS.

This is a painful affection of the bladder of inflammatory origin, characterized by frequent and very painful discharges of small quantities of urine containing pus.

By dieting with Kumysgen, Carnrick's food, lacto-cereal food, or lacto-preparata, the nutrition can be greatly improved and the amount and quality of excretory substances to be discharged with the urine made least irritating. This in itself will often cure many cases without local medication. The cholagogue tablets and Bili-pancreatine pills will be found of great service in this class of affections.

The bladder can be washed out with a mildly astringent or antiseptic solution, as boracic acid, biborate of soda, or salicylic acid.

BLOOD DISEASES.

Under this heading may be included simple anæmia, chlorosis, pernicious anæmia, pseudo-leucocythæmia, leucocythæmia, and Addison's disease, all of which are characterized by a diminution in the hæmoglobin of the blood, while all but the anæmia and chlorosis have positive deficiencies in the number of the corpuscular elements and in the chemical composition of the blood, which up to the present time medical science has not been able to successfully combat ; in all but the first two, death invariably claims the victory.

An attempt, however, should be made to re-establish the nutritive condition, and raise the constructive powers of the system up to a normal standard, which, if successfully accomplished, will result in a cure of the diseased condition.

For this purpose Kumysgen, Carnrick's food, lacto-cereal food, or lacto-preparata will be found most available.

Iron in some form, the corrigent pills, or Blaud's pills. The Bili-pancreatine pill should also be used to aid digestion. Inhalations of oxygen, and the internal administration of arsenic, iodine, tonics, etc., have all been recommended but all alike have failed.

HEART AFFECTIONS.

If an attempt were made to go into the symptoms of all the lesions of the heart, that in itself would make a large pamphlet, if not a full-sized book. But a few practical hints may be the cause of relieving some poor sufferer.

In the management of all this class of diseases a few general principles are in order, and when effectually carried out afford great relief. Remove all digestive disturbances, regulate the action of the heart, and develop a compensatory hypertrophy equal to the cardiac defect.

The digestive disturbance can be perfectly regulated by the use of Kumysgen, lacto-cereal food, lacto-preparata, or the Carnrick's food, either alone or in conjunction with milk, meat, and a moderate allowance of the food-stuffs of vegetable origin. To this should be added the daily use before each meal of one or two Bili-pancreatine pills, or an equivalent of the fluid preparation. By careful dieting with the internal administration of this pill the nutrition in general, and that of the heart in particular, will be improved, and a compensatory hypertrophy developed and maintained.

To slow the heart, increase its working and nutritive capacity, the following tablets will be found most effectual :

"CARDIENE TABLET."

℞

Sulphate of Strychnine	3 grains
Sulphate of Atropine	⅓ grain
Sulphate of Morphine	15 grains

M. Divide into one hundred tablets, of which one to three can be taken three or four times daily.

OBESITY.

The normal percentage of adipose tissue, developed as the result of a perfect physiological condition, is about five per cent. of the total body weight. Every pound added in excess of this percentage is, as a rule, the result of defective oxidization and should be regarded as a pathological process rather than a perfect physiological state. The erroneous idea that fat indicates health, and the attempt to produce fat and over-plump babies has unquestionably been the cause of many an untimely death. For it is well known that fat children and adults do not withstand the inroads of infectious diseases as well as those of a leaner mold. In cases of slow starvation, or with a short allowance of food, an excess of fat within the animal frame may in some instances prolong life longer than otherwise would occur, but in diseased processes this is not found to hold true.

The common causes of obesity are due to small lungs, a diminution in the red corpuscles of the blood for any reason, and the eating excessively of all kinds of food, the taking of alcohol to excess, and the want of outdoor exercise.

The chief symptoms associated with this condition are a troublesome bulkiness of the body and an inability to move about with ease, and considerable dyspnœa.

The treatment consists chiefly in cutting down the quantity of food and regulating the quality in such a manner that the amount taken shall be a little below the physiological standard. Where this plan of dieting is followed, Nature will be compelled to utilize the surplus fat contained in the body to meet the shortage in the foodstuffs taken into the system, and in this way the obesity will gradually be removed. This statement applies particularly to the starches, sugars, and fats.

Kumysgen, lean meat, and a small allowance of bread will give the best results.

While upon this diet, exercise in the open air should be regularly followed, walking or horseback riding being the best, gradually increasing the amount until a normal standard of health is attained.

The cholagogue tablets should be used from time to time, and the mild, strong, or compound and tonic Bilipancreatine pill, as the case may require, to regulate the action of the bowels and aid in perfecting the digestive and assimilative actions.

As there is usually a defective quantity of red blood corpuscles or hæmoglobin in the blood, the corrigent pills should be administered three times daily.

Where the lungs are disproportionately small, as is the case in the short and thick-set built class of people, the oxygenating capacity of the body is, as it were, congenitally defective. In these instances, it will require great medical skill and persistence to establish and maintain a thoroughly good standard of health.

HÆMOGLOBINURIA. *

This is a condition in which dark-colored urine, which contains hæmoglobin, with but few, if any, blood corpuscles, is excreted by the kidneys.

Its causes are muscular exertion, mental excitement, malaria, syphilis, rheumatism, malignant and septic fevers, putrid and typhus fevers, purpura, scurvy, phthisis, etc. After the use or poisoning by napthol, pyrogallic acid, arsenuretted hydrogen, carbon anhydride, chlorate of potash, etc.

The theory of Ponfick appears quite rational, viz., that a condition of hæmoglobinuria is produced by, or is attributable to, a destruction of the red blood discs ; that this condition is divisible into three forms, dependent upon intensity, and in which the corpuscular débris is disposed of in as many different ways. In the *first* or mildest form the *detritus* accumulates in the spleen, which causes an enlargement of that organ.

In the *second* or intermediate degree the liver excretes an excessive quantity of bile (hypercholia), and in this way relieves the system.

In the *third* or severest form the corpuscular débris is excreted from the plasma of the blood by the epithelium of the kidneys, and gives rise to the condition known as hæmoglobinuria.

The enumerated causes certainly point in this direction.

If this theory be true, it offers a very rational and satisfactory explanation for the unusually high colored urine in pneumonia, and the severe forms of disease. It would appear, therefore, that a partial hæmoglobinuria resulting from the destruction of the red blood discs is the true explanation of the high-colored urine. The *hepatic* portion of this theory also affords a tangible reason for the icterus frequently developed in this class of diseases.

The *splenic* portion of the theory explains the universal enlargement, and the metamorphotic changes generally met with in connection with that organ.

---

* Porter: " Renal Diseases." William Wood & Co., 1887, p. 106.

The *renal* portion explains the development of the peculiar urine, and accounts for the precise changes in color with every fluctuation of the acute disease.

The attacks are usually paroxysmal except in connection with the acute infectious diseases, and commence with a chill, rise in temperature, and closely resemble a malarial attack.

During the paroxysm, or early in its subsidence, the urine contains albumen ; later, the urine passed is dark in color, resembling port wine, or even of a deeper hue. In a few hours the urine becomes normal, or it may remain abnormal for days or weeks.

Examination of the urine shows albumen and hæmoglobin as the only abnormality. No blood corpuscles are found to account for the bloody urine. Occasionally a hyaline cast is found.

This disease is quite frequently seen in the south and west, and south-western sections of the United States ; occasionally in all parts of the country.

Dieting is important. At first, Kumysgen, lacto-cereal food, lacto-preparata, or Carnrick's food will be found the most available ; then milk, eggs, lean meat, and a well regulated mixed diet.

Exposure to cold must be avoided, and the body warmly clothed by woolens worn next to the skin, and plenty of out-of-door exercise.

The use of the cholagogue tablet and the Bili-pancreatine pills will aid the digestion and relieve the hepatic portion of the disease.

The corrigent pills and tonics must be given to aid the system in replacing the loss of hæmoglobin.

Remove the cause if possible. If the disease is malarial or syphilitic in origin, anti-periodic treatment with quinine for the malarial cases, and anti-syphilitic measures for those of specific origin, will result in complete cures.

PART II.

THE PHYSIOLOGY OF DIGESTION AND ASSIMILATION.

In this section a little more complete but brief explanation is given for many of the unqualified assertions made necessary by the brevity used in the preceding part.

Now that our chemical and physiological knowledge has become more exact and rendered practically applicable in the study of the composition of our so-called food-stuffs, and with their actions upon the system intelligently appreciated, it naturally becomes essential to note a few of these general laws and the manner in which they are tending to modify and simplify our conception of the true utility of the nutritive agents at our command.

In the past, most diet lists appear to have been constructed upon the varied experiences of a number of well-known observers, but without bringing each item used down to a chemical and physiological test before deciding upon its proper position and usefulness as a food-stuff.

When an attempt is made to find a rational explanation for the rated values of many articles of food in common use, no better ground for their position is found than a most indefinite generalization of hearsay recommendations.

Modern science and chemical investigation, however, has largely changed all this and quite conclusively demonstrated the concise composition of the proximate principles which enter into the formation of the food-stuffs and also shown what their effects upon the functions of the body are when they are introduced into the vascular system of the human economy.

The quantitative composition of the food-stuffs and the atomic construction of the proximate principles entering into them and their actions, together with the results obtained from each substance while passing through the system, are at the present time quite accurately understood.

The comparative composition and the respective merits of the different food-stuffs are much more thoroughly appreciated than formerly, and it is now well known which substances have a high and which a low nutritive value. It has also been clearly demonstrated which of the proximate principles have heating and energy yielding, lubrication and rotundity forming properties only ; which are essential constituents and enter into the chemical and histological construction of all the fluids, ferments, tissues, and glandular structures of the body, and which act simply by their mechanical presence.

The physiological actions of the food elements, the reasons and chemical laws that govern the different kinds, classes, and conditions can be concisely stated and as positively demonstrated.

The best argument of all, however, for the utility of special classes and kinds of food-stuffs is found in well-established clinical facts such as these, that when these distinct chemical and physiological laws are intelligently applied in the proper selection and administration of our food products both in health and disease, results are developed almost at will and with a wonderful degree of uniformity and exactness.

Results are obtained that could not be secured before these underlying and fundamental principles were appreciated and brought into action.

The three classes of proximate principles that are necessary to be understood in the intelligent study of the food-stuffs, and in the selection of the most efficacious diet in disease are best divided into three distinct divisions ; the inorganic, the CHO, and the CHNOS compounds, or a first, second, and third class.

First. The inorganic substances, such as water, the phosphates, chlorides, carbonates, sulphates, etc., etc. These chemical compounds all enter the body under their own form, either alone or in combination with the other two classes. They are not oxidized or split up within the system to enter into the chemical formation of other compounds, but are united mechanically with the proteid group, in fact, their whole action is, as it were, mechanical. After having served their purpose to the body they pass out of the system with the excretions absolutely unchanged in their composition.

All medicinal compounds of a corresponding compo sition and nature probably act in a similar mechanical manner.

Second. The CHO substances which have for their chemical composition the elements carbon, hydrogen, and oxygen only as fat, sugar, and starch. These substances are all oxidized or split up within the system, yielding heat, energy, lubrication, and rotundity only, and are finally eliminated from the body as carbon dioxide and water.

The medicinal agents of like chemical construction probably are oxidized and broken up to yield their effects by a similar cycle of changes.

Third. The CHNOS substances or those which have for their chemical composition the elements carbon, hydrogen, nitrogen, oxygen, and sulphur. The common representatives of this group are called proteids, or the albuminous parts of milk, eggs, meats of all kinds, which chemically and histologically include fish, lobsters, crabs, turtles, oysters, clams, etc., also poultry and game. The nitrogenous or albuminous parts of all plant life, which is now commonly called vegetable proteid, is included in this class.

All these nitrogenous substances, irrespective of specific names, are somewhat slowly oxidized or split up within the system and are absolutely essential to form the different constituents of all the fluids, tissues, glands, and ferments of the body, being united mechanically in varying proportions with water and the mineral salts.

When these proteid bodies are normally transformed, their excrementitious products are urea, uric acid, kreatinine, carbon dioxide, and water in certain and definite proportions.

If for any reason there is an abnormal transformation along this line of proteid metabolism, the relative quantity of urea falls and the uric acid rises. By closely studying these urinary changes and intelligently interpreting them, there is furnished an almost exact key to the perfection or imperfection of the oxidization processes and the nutritive condition of the body. By this method of study it is positively known whether the food-stuffs are absorbed and properly utilized by the system or not.

Another important phenomena to be remembered in connection with the oxidization of the proteid substances is the fact that a disturbance in their anabolism not only changes the relative proportions between the urea and the uric acid, but tends to develop an almost unlimited number of katabolins, some of which are perfectly inert, while others are as toxic and dangerous to life as the well-known cyanide compound prussic acid.

The action of the CHNO medicinal agents can be explained largely upon the same principles and chemical laws that govern the usefulness of the proteid bodies.

With an intelligent conception of these three classes of proximate principles and what results are obtained for the system by their perfect and what by their imperfect oxidization, a comparative table of the common substances rated as food-stuffs is instructive. This table subdivides each kind of food product into its three distinct classes of principles. The inorganic compounds, however, are subdivided into water and the inorganic salts, so that their true position may be more clearly elucidated and the whole subject made plainer.

TABLE I.

COMPARATIVE TABLE OF FOOD-STUFFS.

Kind of Food.	Water, $H_2O.$	Proteids, or CHNOS.	Starch, Sugar and Cellulose or CHO.	Fat or CHO.	Mineral Salts.
Human Milk . .	88.28	3.41	4.62	3.48	0.21
Cow's Milk . . .	86.23	3.73	4.93	4.50	0.60
Skimmed Milk . .	88.00	4.00	5.04	1.80	0.80
Buttermilk . . .	88.00	4.10	6.40	0.70	0.80
Cream	66.00	2.70	2.80	26.70	1.80
Cheese	41.84	29.23	. . . .	23.84	5.09
Eggs	69.05	15.58	. . . .	13.96	1.41
Average Meat . .	65.56	17.51	. . . .	13.16	3.77
Fat Meat . . .	54.22	15.99	. . . .	28.22	1.57
Lean Meat . . .	74.44	19.77	. . . .	2.56	3.23
Average of Fish .	75.57	16.98	. . . .	6.20	1.25
Butter	11.70	0.50	0.50	87.00	0.30
Bread	36.00	7.50	53.85	1.15	1.50
Potatoes	74.50	2.25	21.92	0.15	1.18
Lentils	12.51	24.81	58.36	1.85	2.47
Beans	11.75	24.31	58.03	2.54	3.37
Peas	14.93	23.00	57.80	1.86	2.41
Wheat Flour . .	12.46	14.66	67.62	1.93	3.33
Rye	13.97	14.27	66.91	2.25	2.60
Barley	13.80	12.96	67.18	2.76	3.10
Oatmeal	12.05	12.15	67.00	6.55	2.25
Corn	14.80	12.50	62.65	8.80	1.25
Millet	13.14	12.35	68.35	3.60	2.30
Rice	15.14	7.47	75.69	0.80	0.90
Green Vegetables .	88.00	2.50	3.25	1.75	4.50
Alcohol (?) . . .	0.75	. . . .	99.25	. . . .	. . . .

Having in this brief manner outlined the composition
of the food-stuffs and intimated at the same time the
absolute necessity of understanding thoroughly the chemical and physiological laws that control their usefulness
within the system, it becomes possible to advance a step
further and state the quantities necessary for the most
perfect nutritive condition.

It also shows clearly how, by indulging too freely in
any kind of food or by an unwise selection of the various
kinds of proximate principles, the digestive system is
constantly overtaxed, assimilation imperfectly effected and
a host of diseased conditions developed. These abnor-

malties are brought about in the most insidious and often almost inappreciable manner, until, in some instances, well-defined symptoms are established by which a distinct name can be applied to the condition before attention is attracted to the malady. In a much larger percentage of the cases, however, the symptoms presented are so vague and changeable that the most learned specialist cannot possibly name the condition and sharply define the abnormality so that it can be differentiated from many other states of a similar nature. Yet it is perfectly clear to every one, patient and practitioner, that something is decidedly wrong with the physiological mechanism of the system.

Briefly stated, it may be assumed that the following table, No. II., gives a pretty close and satisfactory basis how the first and second classes of proximate principles should be arranged as to the relative proportions needed of each, from bare subsistence up to the largest amount of mental and physical work.

TABLE II.

COMPARATIVE DIET TABLE.

Diet.	Proteid or CHNOS.		Fat and Sugar, CHO.	
Bare Subsistence . .	100 pts. or 4.00 oz.		100 pts. or 4.00 oz.	
Moderate Exercise .	180 "	8.85 "	161 "	6.56 "
Active Labor . . .	232 "	9.46 "	171 "	6.90 "
Forced Work . . .	242 "	9.86 "	189 "	7.50 "

Before advancing any further in this physiological problem of food and nutrition, it must be admitted that the oxygenating capacity of the system is a limited one — but, fortunately for the human race, it has a moderately wide margin.

There frequently comes a time, however, when this margin is exceeded, which is usually brought about by eating too large quantities of all kinds of food or too freely of the CHO classes of food-stuffs — as fat, sugar, and starches — or of both. As a natural sequence one of three things of necessity follows,

First. The respirations and circulation must be increased to supply more oxygen or the food-stuffs will be imperfectly oxidized. But Nature has set a limit upon the actions of the heart and lungs so that complete relief cannot be granted in this manner,

Second. The red blood corpuscles must be increased in number or empowered to carry more oxygen, or the absorbed food-stuffs will be imperfectly oxidized. But here again Nature has set a limit upon the number and carrying capacity of these anatomical bodies so that relief in this direction is wanting.

Third. The super-abundance of food-stuffs absorbed must be incompletely oxidized because the system has no means by which the extra amount of oxygen required can be furnished. This statement applies with special force to the proteid bodies on account of well-established chemical laws, which show that the CHO elements are quickly and completely transformed under all circumstances while the CHNOS are only perfectly transformed when everything is most favorable.

The CHO compounds, as fat, sugar, and starch, are rapidly and easily oxidized, consequently they are the first elements to be changed, and they are also completely transformed into their final products ; this tends to leave a deficient quantity of oxygen to act upon and accomplish the more difficult task of carrying the nitrogenous compounds through their cycle of change and finally into perfect excrementitious substances.

This defective supply of oxygen disturbs the perfect metabolism of the proteid bodies and produces an unlimited number of katabolins and furnishes a rational explanation for many, if not all, of the pathological conditions and symptoms that have to be treated. At least it is fair to assume that so long as the anabolic processes of the body are perfectly effected, no pathological lesion or abnormal symptom can be developed.

Keeping constantly in mind the table indicating the relative proportions existing between the proteids and the CHO compounds, or the fats, sugars, and starches, and studying a little more closely the composition and comparative merits of the various food products, much valuable information is brought to light.

First. It is found almost impossible to arrange a mixed or vegetable diet so as to obtain the requisite amount of proteid elements without at the same time taking more than the needed quantity of the CHO compounds, that is, without introducing more than can be safely utilized or oxidized.

Second. It is found that with an exclusive meat diet composed of the ordinary average meat almost the exact quantities of both the CHNOS and CHO compounds can be obtained from bare subsistence up to that for forced work.

Taking four ounces of pure proteid matter as the standard amount required in twenty-four hours to perfectly maintain the constructive forces of the system, the following tables are quite instructive, viz. :

TABLE III.

	Water or H_2O.	Proteid or CHNOS	Fat, Sugar, CHO.	Salts.	Totals.
20 oz. Lean Meat .	18.46 oz.	3.96 oz.	0.52 oz.	0.64 oz.	19.98 oz.
112 " Milk . . .	96.54 "	4.04 "	10.53 "	0.68 "	111.83 "
16 " Beans . .	1.88 "	3.89 "	9.28 "	0.54 "	15.59 "
60 " Bread . .	21.60 "	4.13 "	35.05 "	0.72 "	59.48 "
170 " Potatoes . .	125.80 "	3.82 "	37.44 "	2.22 "	169.28 "

TABLE IV.

DIET COMPOSED OF MEAT AND BREAD.

	Water or H_2O.	Proteid, CHNOS.	Fat, Sugar, CHO.	Salts.	Totals.
10 oz. Lean Meat .	7.44 oz.	1.99 oz.	0.26 oz.	.32 oz.	10.01 oz.
30 " Bread . . .	16.80 "	2.07 "	10.20 "	.36 "	29.43 "
	24.24 oz.	4.06 oz.	10.46 oz.	.68 oz.	39.44 oz.

TABLE V.

DIET COMPOSED OF LEAN MEAT AND POTATOES.

	Water or H₂O.	Proteid or CHNOS.	Fat, Sugar, CHO.	Salts.
10 oz. Lean Meat . .	7.44 oz.	1.99 oz.	0.26 oz.	0.32 oz.
80 " Potatoes . . .	59.60 "	1.80 "	17.54 "	0.95 "
	67.04 oz.	3.79 oz.	17.80 oz.	1.27 oz.

TABLE VI.

DIET COMPOSED OF MEAT AND LENTILS.

	Water, H₂O.	Proteid, CHNOS.	Fat, Sugar, CHO.	Salts.
10 oz. of Lean Meat . .	7.44 oz.	1.99 oz.	0.26 oz.	0.32 oz.
8 " Lentils . . .	1.00 "	1.99 "	4.82 "	0.20 "
	8.44 oz.	3.98 oz.	5.08 oz.	0.52 oz.

TABLE VII.

DIET COMPOSED EXCLUSIVELY OF ORDINARY MEAT.

	Water, H₂O.	Proteid or CHNOS.	Fat or CHO.	Salts.
100 oz. Ordinary Meat .	65.56 oz.	17.51 oz.	13.61 oz.	3.77 oz.
23 " " " .	14.30 "	4.01 "	3.02 "	0.86 "
44 " " " .	29.56 "	7.62 "	5.79 "	1.65 "
52 " " " .	34.06 "	9.02 "	6.81 "	1.92 "
56 " " " .	36.68 "	9.96 "	7.34 "	2.02 "

Examples of this kind might be multiplied almost ad infinitum. With all of the tables, however, excepting Table No. VI., there is clearly shown a larger quantity of the CHO compounds than is found of proteid elements. This shows that with almost all kinds of food-stuffs and especially when taken in excessive quantities the system is liable to receive a superabundance of the CHO substances. The ease with which the requisite amounts of proteid matter can be rightly adjusted to meet the demands of the system is clearly demonstrated. These tables just as clearly illustrate that it is almost impossible to arrange any form of the mixed food-stuffs in such a manner that the system will not be constantly supercharged with the stimulating and non-nutritious compounds of CHO construction.

TABLE VIII.

This comprehensive comparative diet table, compiled and used by Prof. William H. Porter, of the New York Post-Graduate Medical School, has been worked out upon the atomic basis of the proximate principles, which enter into the construction of the ordinary food-stuffs.

It proves quite conclusively that Professor Porter's animal diet yields all that can be obtained by the use of a mixed diet containing the three elements — proteid, fat, and carbohydrate.

In fact, in the proportions as here given, it calls for the use of a little more oxygen than the mixed diet based upon the proportions given by Moleschott; it also yields a little more carbon dioxide and water.

When it is remembered, however, that in the egg and the ordinary run of good meat, the proteid element is always a little more abundant than the fat, this excess of oxygen used — when taking an ordinary animal diet — will not be required, and the increased amount of carbon dioxide and water will not be produced, but the total results in excrementitious products cast off and the amount of heat and so-called energy evolved will come so very close to the amounts obtained by using Moleschott's mixed diet, that the two are practically the same.

The conclusion, therefore, is that the relative proportions of these two elements, proteid and fat, as commonly found in eggs, meat, and fish, come so nearly to the required physiological demands of the system that, in this class of food-stuffs, there is found an almost perfect standard for diet. By adding a very small allowance of bread and butter, it becomes absolutely perfect.

These chemical facts, based upon the atomicity of the food elements, explain the higher nutritive vitality devel-

TABLE VIII.

COMPARATIVE DIETS AS ARRANGED BY WILLIAM HENRY PORTER, M.D.

Porter's Meat or Proteid and Fat Diet.

		Amount of O used	Urea	Uric Acid	Kreatinine	Carbon Dioxide	Water	Sulphuric Acid
130 grammes 4.5 oz. Proteid	$C_{72}H_{112}N_{18}O_{22}S$	$+18,070(O) =$	$520(CH_4N_2O)+130(C_5H_4N_4O_3)+260(C_4H_7N_3O)+$			$7,150(CO_2)+$	$4,040(H_2O)+130(H_2SO_4)$	
130 grammes 4.5 oz. Fat	$C_{88}H_{182}$	O_2	$+20,345(O) =$				$7,150(CO_2)+$	$6,825(H_2O)$	
		$38,415(O) =$	$520(CH_4N_2O)+130(C_5H_4N_4O_3)+260(C_4H_7N_3O)+14,300(CO_2)+11,765(H_2O)+130(H_2SO_4)$					

Working Power, 91,763 foot pounds.

Moleschott's Proteid Fat and Carbohydrate Diet.

130 grammes 4.5 oz. Proteid	$C_{72}H_{112}N_{17}O_{22}S$	$+16,070(O) =$	$520(CH_4N_2O)+130(C_5H_4N_4O_3)+260(C_4H_7N_3O)+$			$7,150(CO_2)+$	$4,040(H_2O)+130(H_2SO_4)$	
90 grammes 3.1 oz. Fat	$C_{88}H_{108}$	O_2	$+14,085(O) =$				$4,950(CO_2)+$	$4,725(H_2O)$	
330 grammes 11.6 oz. Carbohydrate	C_6H_{12}	O_6	$+3,960(O) =$				$1,980(CO_2)+$	$1,930(H_2O)$	
		$36,115(O) =$	$520(CH_4N_2O)+130(C_5H_4N_4O_3)+260(C_4H_7N_3O)+14,080(CO_2)+11,645(H_2O)+130(H_2SO_4)$					

Total difference between Porter and Moleschott Diet, 2,300(O)   220(CO₂)+ 120(H₂O)

Purely Vegetable Diet of Beans.

260 grammes 9.0 oz. Proteid	$C_{72}H_{112}N_{18}O_{22}S$	$+36,140(O) =$	$1,040(CH_4N_2O)+260(C_5H_4N_4O_3)+520(C_4H_7N_3O)+14,300(CO_2)+9,880(H_2O)+260(H_2SO_4)$					
22 grammes 7.7 oz. Fat	$C_{88}H_{108}$	O_2	$+3,443(O) =$				$1,210(CO_2)+$	$1,155(H_2O)$	
634 grammes 22.3 oz. Carbohydrate	C_6H_{12}	O_6	$+7,608(O) =$				$3,804(CO_2)+$	$3,804(H_2O)$	
		$47,191(O) =$	$1,040(CH_4N_2O)+260(C_5H_4N_4O_3)+520(C_4H_7N_3O)+19,314(CO_2)+14,839(H_2O)+260(H_2SO_4)$					

Working Power, 102,587 foot pounds.

Difference in the amount of oxygen used between the Vegetable and Moleschott's Diet, and between the Vegetable and Porter's Diet; also the differences in the amount of excretory products to be eliminated.

Total difference between Vegetable and Moleschott's Diet

$$\begin{cases} 47,191(O) = 1,040(CH_4N_2O)+260(C_5H_4N_4O_3)+520(C_4H_7N_3O)+19,314(CO_2)+14,839(H_2O)+260(H_2SO_4) \\ 36,115(O) = 520(CH_4N_2O)+130(C_5H_4N_4O_3)+260(C_4H_7N_3O)+14,080(CO_2)+11,765(H_2O)+130(H_2SO_4) \\ \overline{11,076(O)} = 520(CH_4N_2O)+130(C_5H_4N_4O_3)+260(C_4H_7N_3O)+5,234(CO_2)+3,194(H_2O)+130(H_2SO_4) \end{cases}$$

Total difference between a Vegetable and Porter's Diet

$$\begin{cases} 47,191(O) = 1,040(CH_4N_2O)+260(C_5H_4N_4O_3)+520(C_4H_7N_3O)+19,314(CO_2)+14,839(H_2O)+260(H_2SO_4) \\ 38,415(O) = 520(CH_4N_2O)+130(C_5H_4N_4O_3)+260(C_4H_7N_3O)+14,300(CO_2)+11,765(H_2O)+130(H_2SO_4) \\ \overline{8,770(O)} = 520(CH_4N_2O)+130(C_5H_4N_4O_3)+260(C_4H_7N_3O)+5,014(CO_2)+3,074(H_2O)+130(H_2SO_4) \end{cases}$$

Amount of Work Performed upon Porter's Proteid and Fat Diet.

130 grammes of Proteid	$\times 1,812 =$	235,560 kilogramme-metres, or	32,567 foot pounds.
130 " " Fat	$\times 3,841 =$	499,330 "	69,035 "
		734,890 kilogramme-metres, or	101,602 foot pounds.

Amount of Work Performed upon Moleschott's Proteid Fat and Carbohydrate Diet.

130 grammes of Proteid	$\times 1,812 =$	235,560 kilogramme-metres, or	32,567 foot pounds.
90 " " Fat	$\times 3,841 =$	346,690 "	47,793 "
330 " " Carbohydrate	$\times 1,637 =$	540,810 "	74,770 "
		1,128,060 kilogramme-metres, or	155,960 foot pounds.

The same as above but with Carbohydrate error corrected.

130 grammes of Proteid	$\times 1,812 =$	235,560 kilogramme-metres, or	32,567 foot pounds.
90 " " Fat	$\times 3,841 =$	345,690 "	47,793 "
330 " " Carbohydrate $\times$ 294	$=$	97,020 "	13,413 "
		678,270 kilogramme-metres, or	93,773 foot pounds.

Amount of Work Performed upon a purely Vegetable Diet of Beans.

260 grammes of Proteid	$\times 1,812 =$	471,120 kilogramme-metres, or	65,134 foot pounds.
22 " " Fat	$\times 3,841 =$	84,502 "	11,683 "
634 " " Carbohydrate $\times$ 294	$=$	186,396 "	25,770 "
		18 kilogramme-metres, or	102,587 foot pounds.

Work accomplished on Moleschott's diet	155,960 foot pounds.
" " Porter's "	101,602 "
Total difference in favor of Moleschott's,	54,358 "

Work accomplished on Porter's diet	101,602 foot pounds.
" " Moleschott's corrected diet,	93,773 "
Total difference in favor of Porter's	7,829 "

Work accomplished on Vegetable diet	102,587 foot pounds.
" " Porter's "	101,602 "
Total difference in favor of Vegetable	985 "

Work accomplished on Vegetable diet	102,587 foot pounds.
" " Moleschott's corrected diet,	93,773 "
Total in favor of the Vegetable diet	8,814 "

oped in the carnivora as compared with the herbivora and vegetable feeding classes.

Again, in diseased conditions where the nutritive powers are severely overtaxed, the proteid and fat diet is especially serviceable, for by its use the expenditure of vital force in transforming the food-stuffs is kept at the lowest possible standard. Because, in the use of animal fat to the exclusion of the carbohydrates, the system is spared the necessity of laying out force and oxygen to convert the starch and various sugar elements into a diffusible glucose, and then into an alcoholic-like compound before they can be utilized by the animal economy for the production of heat and the so-called energy, which is finally computed in foot pounds of work accomplished.

This great saving in vital force by the exclusive use of fat — to supply the CHO elements necessary to produce the heat and energy required — is unquestionably the exact factor that enables the system to effect the cure in all the pathological conditions, which otherwise could not be carried on to a complete recovery.

These same laws make Kumysgen one of the most valuable food products ever produced, because it has been found, that only about one-half of the fat contained in milk is capable of being absorbed, and with the lactose converted into an alcoholic compound there is developed in Kumyss or Kumysgen, particularly in the latter, a partially predigested food-stuff which contains about equal quantities of proteid and fat in a state to be readily absorbed. This then corresponds exactly with the requirements found in Professor Porter's table, which consists of only proteid and fat.

Practical experience has long since taught that this form of dieting was the only kind available in connection with the successful treatment of the acute diseases.

This table is further a demonstration and confirmation from a chemical and physiological standpoint, of what has been so often repeated in a clinical way, that upon this purely proteid and fat diet, together with the administration of suitable medicinal agents, the most aggravated forms of digestive disturbances can be quickly removed, nutrition improved, and a healthy standard permanently re-established. They show conclusively that this form of animal diet yields the largest working power to the system for a given amount of food taken, and a similar amount of oxygen used to carry the food substances through their anabolic cycle of changes, and finally form and discharge from the body the resulting excrementitious products.

At this point, however, it may be well to mention that the standard amount of proteid matter taken, in the construction of all these tables, was 130 grammes — 4.5 ounces. Moleschott's original diet-table contained only 120 grammes or (4.2 ounces), but as almost all observers agree quite closely as to the amount of proteid material necessary to be used, and also as to the results obtained from its oxidization, the same quantity was used in all instances that a more exact comparison might be established. The chief difference of dispute, however, is in relation to the relative value of the fats and carbohydrates, and particularly in reference to the latter compounds.

In trying to develop out of a purely vegetable diet, anything like the same amount of working power for the system that is obtainable by the use of Porter's or Moleschott's diet, almost double the amount of proteid had to be taken with the proportionate rise in the fat and starch as is contained in the vegetable chosen.

To produce the same amount of work by using a vegetable diet necessitates the outlay of a much larger amount of oxygen, and the production and handling by the glandular structures of the body of an excessive amount of the nitrogenous excrementitious elements. These facts illustrate quite conclusively the manner in which the damage to the system is brought about by indulging too freely, or living exclusively upon a cereal or vegetable compound.

The vegetable proteid in these tables is further given an undue advantage, to which it is not justly entitled, by crediting it with the same atomic formula as that possessed by an animal proteid; since the nitrogenous element found in plant-life contains a much larger number of nitrogen atoms, and consequently requires more vital force and oxygen to digest and assimilate it. This naturally decreases rather than improves the nutritive value of the proteid compound of vegetable origin.

An average of a compound fat molecule is taken as the working standard in all these tables.

Attention is also directed to a probable error in the rating of the heat-producing power of the carbohydrate. It is commonly stated, that the comparative oxygenating capacity of a carbohydrate and fat is as one to two and one-half, but by their chemical atomicities, it is as one to thirteen, or thirteen and one-half in favor of the fat.

That such an error exists in the computations in Moleschott's standard is sustained by a comparative study of the atomicities of the food-stuffs used in both Porter's and Moleschott's diet tables, and of the amount of oxygen required for complete oxidization in both instances. In the former, or Porter's proteid and fat diet table, a little more oxygen is needed than is necessary in Moleschott's mixed diet, yet it is claimed that in the latter instance 393,170 kilogramme-metres or 54,358 more foot pounds of work is produced. This, however, is directly opposed by the smaller quantity of oxygen used in the oxidization processes. When this error in work, produced out of the carbohydrates in Moleschott's diet, is corrected in accordance with the difference in atomicity and the amount of oxygen used between the fat molecule and the carbohydrate molecule represented as glucose, and a computation is made in accord with the correction, a slight difference in work produced when living on a Moleschott's or Porter's diet, is found to exist. The increase in work produced, however, is now found to exist in connection with Porter's diet and is in accord with the larger amount of oxygen used, which makes atomicity, oxygen used, and work produced correspond, while the reverse was stated in the calculations formerly made in connection with Moleschott's diet.

If this error be true, as it appears to be, the profession have been sadly misguided in all their attempts in the construction of diet tables starting with Moleschott as their standard.

On the other side, if these chemical and physiological laws be true, as based upon the atomicity of the proximate principles, by carefully considering the percentage composition of each food product to be used, exact results can be obtained.

Another point to which attention is called by Dr. Porter is this, that the factors 1.812 and 3.841, which are used in computing the kilogramme-metres in Table VIII., are taken from Frankland — Philosophical Magazine XXXII., and are those which are generally quoted in all scientific works upon physiological chemistry and upon diet.

In studying the proximate principles, however, by the atomicities, and considering the amount of oxygen required to completely transform a fat molecule into its final products of excretion water and carbon dioxide and a proteid molecule into its final products of excretion — urea, uric acid, kreatinine, carbon dioxide, water, etc.— it is found that only eighteen (18) more oxygen elements are used in the complete oxidization of the fat than in that of the proteid molecule. The computed amount of work performed by the oxidization of the fat molecule is found to be 530 foot pounds as compared to 250 foot pounds for the complete oxidization of the proteid molecule. This makes the eighteen (18) more elements of oxygen used in transforming the fat molecule result in the production of 280 more foot pounds of work than is obtained from the eighteen less used in the proteid.

From this a decided discrepancy is quite evident between the results obtainable by former calculations and those based upon our modern chemical atomicities.

However, for an illustrative and comparative study of the working power obtainable from the use of the various food-stuffs, this table is still of great value, as the same figures are used in each and all the calculations.

As these same factors, 1.812 and 3.841, appear in all the modern scientific works, they were retained in the arrangement of this table, but not without appreciating and calling attention to this discrepancy when the computation is based upon the atomicities of the food elements used, the amount of oxygen required, and the results obtained.

Again, it must be remembered that the proteids are not directly transformed into their final products, but undergo a series of intermediate changes, all of which require the use of oxygen and must of necessity yield more or less heat and energy, so that all our estimates are approximate.

When upon Moleschott's diet with the proteid substances raised to the common standard of 130 grammes and the carbohydrates rated in accord with the correction previously noted, it requires 36,115 oxygen elements to produce 678,270 kilogramme-metres or 93,773 foot pounds of work.

When upon Porter's diet of proteid and fat, it requires 38,415 oxygen elements to produce 734,890 kilogramme-metres or 101,602 foot pounds of work.

When upon a purely vegetable diet that will yield anything like the requisite amount of work that can be obtained by using Moleschott's or Porter's diet, it requires 47,191 oxygen elements to produce 742,018 kilogramme-metres or 102,587 foot pounds of work.

To obtain the 63,748 more kilogramme-metres or 8,814 foot pounds of work out of the vegetable diet as compared with Moleschott's diet, it requires the expenditure of 11,076 more oxygen elements.

To obtain the 7,128 more kilogramme-metres or 985 foot pounds of work out of the vegetable diet as compared with Porter's diet, it requires the expenditure of 8,776 more oxygen elements. The vegetable diet in both instances yielding an excessive amount of nitrogenous excretory matter, carbon dioxide, and water.

A careful study of Table II. and VII., and Porter's diet in Table VIII., proves beyond a question of doubt that upon an exclusive diet of our ordinary average meat alone very nearly the required proportions of the proteids or CHNOS compounds and of the fat or CHO element can be established.

The only defect in the perfection of Table VII. and VIII. is found in the saline column, which contains much more mineral matter than perfect physiological laws indicate are required. This excess in saline or inorganic compounds, however, appears to be true in all kinds of food products — that is, if the proportion of salts in the milk is taken as the guide for a working basis. The reason for looking upon the amount of salts in the milk as the guide to the maximum quantity required is based upon the fact that during the infant period of life, where milk forms the only source of food supply, bone formation is most rapidly progressing, and the amount of mineral matter needed by the system is at its height and much larger than at any other period of life. The bones continue to grow and become fully and perfectly developed with the ordinary quantity of mineral matter contained in the milk.

Physiology also teaches that a little less than one ounce of mineral salts are required daily by the system, but in all the tables given, except the one containing milk alone, the amount of salts is fully up to or more than an ounce.

The only great objection that can be raised to an exclusive meat diet is the lack of variety, but that is quite easily adjusted by varying the kinds of meat used. The perfection of the proportionate composition of the proximate principles when using a meat diet, the smaller liability to imbibe an excessive quantity of any one kind and the little danger that there is of taking an excess of the CHO or stimulating and non-nutritious compounds, clearly establishes the fact that in meat we approach the nearest to an ideal food.

If attention is turned for a single moment to the lower orders of the animal kingdom, it is quite apparent that the most supple and intensely powerful organisms are found among the carnivora only. This tends to substantiate the high utility of the meat diet. Another interesting point is the almost universal absence of tuberculosis among meat-eating animals, while the vegetable-feeding class are specially prone to suffer from this fatal malady.

Again, the milk which is so generally considered as being fully equal to all the demands of the system, and especially so during the first few months of infant life, might be brought forward as proof positive and clearly illustrating the fact that nature calls for an excess of the CHO elements, because in the composition of the milk it is found that the CHO substances are about twice as abundant as the proteid or CHNOS elements. When these facts are examined a little more closely and scientifically, it is found that the pancreatic gland and its ferment-forming bodies are imperfecty developed at this period of life. Consequently, the fat, if emulsified and rendered capable of being absorbed by the lacteals of the villi, must have this transformation effected almost exclusively by the biliary fluid alone. It is further taught that the biliary secretion acts but little, if at all, upon vegetable fats and that it has the power to effectually emulsify only about one-half of the total quantity of animal fat introduced into the alimentary canal. This being true, fifty per cent. of the fat contained in the milk, together with the bile constantly flowing into the alimentary tract, is unquestionably utilized by the system as a natural laxative principle, and is undoubtedly the chief method by which nature effectually maintains the regular movements of the bowels and produces the daily evacuations so characteristic of a perfectly healthy infant.

The proportionately larger size of the liver in a child as compared with an adult also points to the fundamental importance of the hepatic gland and its secretion as a necessary agent of prime importance in the infant; the large size of the liver compared with other organs also indicates its great importance during adult life.

How much of the lactose — which is the form of sugar introduced in the milk — is inverted into glucose and rendered capable of being absorbed and utilized is an open question. In fact, there is no very reliable data upon this important point, but what is to be found upon the subject indicates quite positively that a considerable quantity of the lactose is not changed so as to be utilized by the system, but passes off with the fæces. Therefore, when the scientific truth is clearly appreciated, it is found that the relative proportion between the CHO and the CHNOS elements contained in the milk and that which can gain access to the vascular channels and be of service to the system is not far from equal in amount the major quantity, perhaps a little on the side of the CHO substances or in favor of the fat and sugar. Then, again, the infant requires a little more of the heat-producing compounds during the first few weeks or months than is needed a little later on or in adult life, because the proportionate amount of energy expended is greater in the infant and child than is the case during the adult period of life. Very early in the infant life there is comparatively little muscular action by which heat and energy can be evolved, while a large amount of heat is needed to maintain a perfect physiological condition, and for a time warmth must be artificially supplied.   These conditions will admit of a little excess of the CHO elements during this period of life, but when the stage of infant muscular activity commences its never-ceasing motion, then the proportionate amount of the proteid substances must be raised and the CHO, or fat and sugar lowered, if the most perfect type of physiological development is to be effected.

Observing clinical phenomena a little more closely, it is quite apparent, as life advances, that milk is not equal to the demands of the system, and a more strongly proteid diet is urgently called for by nature.   Eggs and lean meat must next be added to furnish this much-needed proteid pabulum for the constructive purposes of the animal economy, and out of which alone the most perfect muscles, glands, ferment bodies, and brain tissue can be formed.

By this process of reasoning, it is clearly and well established that even with the commonly supposed typical food-stuff, milk, it is not sufficiently perfect in its composition to thoroughly sustain the nutritive economy under all circumstances, but must have added to it a more liberal proteid pabulum.   It is also clearly demonstrated that a portion of this excessive amount of fat is not taken up by the circulatory or lymphatic system but is used largely by nature as a laxative agent.

Proceeding a step further in the investigation of the clinical facts bearing upon this most interesting subject and there is found quite a common tendency among people at large to add to the nutritive supply of the infant not the most serviceable kind of food-stuffs in the way of an animal proteid of some kind, but on the contrary the more general practice is that of adding a cereal or vegetable compound, — one in which the CHO elements are very greatly in excess of the demands of nature.   Another important point to be remembered in this connection is the well established fact that, although the proteid of vegetable origin, while in quite sufficient quantities, is a much higher nitrogenous compound and, as a rule, is far more difficult of digestion than a proteid body derived from the animal kingdom.

By this method of infant feeding in which an excess of the fat, sugar, and starch or CHO compounds are used, a natural taste and habit of eating food derived largely from

the vegetable kingdom is engendered. The natural
sequence is, that on through life the individual is apt to
continue eating excessively of all kinds of food-stuffs and
particularly those of the CHO and vegetable class. This
poorly nourishes the body; adipose tissue in abundance is
often acquired from the imperfectly transformed food-
products. The appetite increases because the system is
not properly sustained. The individual continues eating
more and more until finally the marginal capacity of the
system for supplying oxygen is reached and passed, digestion is imperfectly effected, and the oxidization powers of
the body exceeded.

The fats and sugars or CHO compounds, by the laws of
chemistry being most easily oxidized, are transformed first
and thus exhaust the oxygenating capacity of the animal
economy and leave, as has been already explained, a deficient supply of oxygen for the more difficult task of oxidiz-
ing the proteid substances. As a result of this condition
uric acid takes the place of urea in the urine. Owing frequently to the excessive amounts of mineral salts contained
in and taken into the system with the vegetable foods —
and in fact with meat if ingested too freely — the true condition may be obscured. For, when these two substances,
uric acid and the mineral salts, reach the urinary channels in
excessive amounts, this powerful organic acid, uric acid,
attacks the salts of soda, potash, lime, etc., and forms
normal and acid salts of the urates and a superabundance
of earthy phosphates. The former being fairly soluble are
held in solution, while the insoluble phosphates tend to
spontaneous precipitation, and leads to the erroneous
impression that there is a phosphatic condition, while the
true and principal disturbance is an incomplete oxidization
of the proteids and an excessive production of uric acid
instead of the more complete product, urea. This mask-
ing of the uric acid condition has often prevented the true
state of the system being discovered. But recent chemical
investigation has clearly explained these cases and placed
the physician in a position to recognize at once the precise
situation.

The necessary sequences of these habits of false and
over-feeding and incomplete oxidization of the proteid

substances is to impair the nutritive integrity of the whole
system, disturb all the physiological functions of the body,
and produce a large number of functional and pathological
conditions, some of which have been specifically named, but
in other instances, the vagueness and variability of the
symptoms produced, precludes all possibility of well-defined
names, by which one condition can be distinguished from
another.

From this line of study, it is clearly demonstrated that
the chief disturbing elements in all these diseased conditions,
by which the symptoms are produced come from the incomplete or imperfect metabolism of the proteid bodies, and
exists in the form of a katabolin of some kind.

As the liver is the chief organ concerned in the transformation of these proteid substances, when they are first
introduced into the system through the entero-hepatic circulation, and also in their anabolic changes on their way
back from the tissues to be finally converted into and
eliminated from the system as urea, uric acid, kreatinine
carbon dioxide, and water, the physiological activity of
the epithelial cells, which constitute this hepatic gland, is
among the first and one of the chief organs in the body to
suffer from prolonged over-feeding. This is evinced prin-
cipally by an alteration in the bile-producing action of the
liver, and as a result the bile poured into the duodenum is
defective both in quality and quantity or the amount secreted
is augmented, but its quality still remains very much
impaired.

At first the deteriorated condition of the quality of the
biliary fluid produces very little apparent disturbance, but,
after a time, on account of the great importance of the
hepatic secretion as a digestive agent, the food-stuffs are
imperfectly transformed in the intestine, and now the nutritive pabulum carried up to the liver for the nutrition of the
body is decreased in quality, less perfectly transformed and
frequently mixed with a considerable quantity of deleteri-
ous material which has been generated within the intestinal
canal, as the result of the defective supply of this important
digestive agent — bile.

Nutrition becomes more and more impaired, and the
pancreatic gland and its ferments deteriorated. In fact,

the same reduction in the nutritive and functional activity holds true for the stomach and throughout the whole body. The primary faulty action, however, is in the liver and biliary fluid, and secondarily, the pancreas and its secretion fail.

To meet these defects the Bili-pancreatine pill has been made and can be strongly recommended both upon clinical and physiological grounds.

Another very pronounced clinical fact to be remembered as substantiating the theory that over-feeding and an excessive use of the CHO compounds is the fundamental cause of most of our diseased processes, when not directly traceable to a specific poison which has been introduced into the system, is the well known law that, in all diseased conditions, the diet must be reduced in quantity and limited almost exclusively to an animal proteid compound if success is to crown the efforts made to sustain and restore an over-taxed system to a normal state.

In fact, in the treatment of the most critical conditions a step in advance even of this is taken, which consists in displacing almost all of the CHO compounds commonly used as food-stuffs, as fat, sugar, and starch, and replacing these by the C_2H_6O compound alcohol. This is done because the alcoholic-forming substances have already been transformed outside of the body and converted into an easily absorbable compound, which, molecule for molecule, is quickly oxidized within the system with a small expenditure of oxygen, and at the same time it yields a very large amount of energy for the force expended in utilizing the alcoholic compound.

In this way the digestive powers are held in abeyance, and the highest type of nutritive activity established out of the proteids administered. While alcohol is used instead of the fat, sugar, and starches, the quantity administered should be kept well within the physiological bounds, and an animal form of proteid ingested in preference to that of vegetable origin, for then the most perfect nutritive standard is obtained, the vital and constructive forces sustained, the diseased condition quickly removed, and perfect health fully established.

This line of argument has not been advanced with the view of ruling the vegetable proteid or the CHO class of

compounds from our dietary lists, but rather to show how
an excessive use in all kinds of food-stuffs and particularly
the CHO and vegetable compounds bring about many of
our abnormal conditions which have so long baffled explanation and resisted treatment.

This illustrates, how by having a thorough knowledge
of our chemical and physiological laws in conjunction with
an abundance of clinical evidence, the precise manner in
which all the abnormal conditions that are developed can
be understood, and at the same time it indicates how easily
an application of these principles with well selected
remedial agents will cure many a case that has long since
been pronounced chronic or incurable.

Having pointed out the great importance of the hepatic
and pancreatic secretions in connection with the digestive
processes, a careful consideration of the biliary fluid and
its functions is essential to clearly elucidate this important
subject, and the secretion derived from the pancreatic gland
must also be studied.

Normal bile as secreted by the hepatic cells is a perfectly clear fluid of neutral reaction, but as it passes along
the bile ducts and has added to it the temporarily detained
bile of the gall-bladder, it obtains a varying quantity of the
mucus secretion produced by that receptacle, after which it
becomes somewhat more opaque and often decidedly tenacious. If the amount of secretion added to it by the gall-
bladder is considerable in quantity, it may, when discharged
into the duodenum, have a slightly alkaline reaction.
But under the most favorable circumstances, the biliary
fluid is neutral in reaction and has a specific gravity ranging from 1010 to 1032. The probability being that when
the bile is not carrying off considerable waste material, but
is simply acting as a normal secretory fluid, its specific
gravity is low, like all the other digestive fluids. When,
however, it is acting largely as an excretory fluid and carrying out of the system a considerable quantity of effete
products, the reverse is true, and it reaches this higher
standard of density, simulating that of the urinary excretion.

The chemical composition of the biliary fluid, like most
of the animal secretions and excretions is, of necessity,
quite complex, and to enter into all its details would carry

us far beyond the scope of this pamphlet. Suffice to say, before enumerating these functions of the bile, that the utility of this fluid is rapidly becoming accurately and scientifically understood. Its paramount importance as an element in the digestion is at once appreciated when we observe the very large amount of work, that has to be accomplished by this fluid and its power to generate active digestive agents. These facts also naturally suggest the absolute necessity of at once removing the old and deteriorated bile before attempting to help Nature to eradicate any and all diseased conditions from the system. It further illustrates the positive demand on the part of the animal organism for a fresh supply of new bile, having a good quality to replace the old and useless bile which has been previously removed.

The functions of the biliary fluids have recently been quite accurately defined by Professor William H. Porter, in his instructive course of lectures,* delivered during the winter of 1890 and 1891, at the New York Post-Graduate Medical School and Hospital, and are as follows :

"*First.*—It has an emulsionizing action, that is, it acts upon neutral fats and minutely subdivides the oil globules so that they can be easily taken up by the epithelial cells and discharged into the lacteals. It does not appear to have the property of splitting fats into their component parts — a fatty acid and glycerine — but, when fully mingled with the biliary fluid, the fat is placed in such a condition that the steapsin of the pancreatic fluid can more easily separate it into a fatty acid and glycerine, and form a soap by the union of the soda and fatty acid."

As these soaps are not capable of being absorbed to any extent, their action together with the glycerine appears to be that of a natural laxative.

"*Second.*—The bile contains a diastatic ferment which has for its action the conversion of starch and glycogen in glucose."

This action is, comparatively speaking, a feeble one, and Prof. Porter, since writing the above, has come to doubt the reversion of the glycogen back to glucose.

---

* *The Medical News*, Philadelphia, January 10 and 24 and February 28, 1891.

"*Third.*—Taken as a whole, the bile may be regarded as a stimulant, producing peristalsis of the intestines, thus tending to favor absorption and prevent constipation."

"*Fourth.*—The bile acids in particular are stimulants to the small muscles contained in the villi, so that by the contraction of these muscles the contents of the lymph-spaces are emptied toward the larger and deeper lymphatics, leaving those in the villi in a position to absorb more. Regarded in this light, the lymph sacs of the villi are little pumps which are constantly pumping fat out of the epithelial cells and discharging it into the deeper lacteals to be passed on to the general circulation."

"*Fifth.*—The moistening of the epithelial cells with bile seems to be necessary for their vital activity, enabling them to take up both glucose and peptones as well as the fat much more rapidly than is the case where they are deprived of a free supply of bile.

"The bile seems also to stimulate a watery flow from the intestinal follicles, which together with its tendency to excite peristalsis forms a free and easy passage of the intestinal contents into and through the colon. By some the bile is spoken of as a lubricant to the colon, which, combined with the other effects, favors free movements and prevents constipation."

"*Sixth.*—The biliary fluid prevents decomposition in the alimentary tract, and may be looked upon as Nature's chief antiseptic. This assertion is sustained by the rapidity with which decomposition occurs whenever the biliary secretion is impaired, totally arrested, or changed in quality. Under these circumstances we notice a tendency to persistent constipation, clay-colored stools, and the distention of the intestines with gas. If the quantity of bile is abundant, but of poor quality, there is a tendency to black, tarry, and burning stools, with some tympanites, and fermentation and incomplete digestion of the food-stuffs."

"*Seventh.*—I think we are justified in assuming that the bile and its contained ingredients aid very materially in the final solution of the albuminoids, and their conversion into diffusible peptones. This view is entertained after a careful study of a large number of cases which has clearly shown that unless the old bile and that of poor quality be expelled from the system, and its place taken by new bile, and that of a better quality, the digestive disturbances cannot be overcome ; but that so soon as this change is effected, digestion and assimilation become perfect and nutrition is carried on normally. This is conclusively proven by the fact that the products of oxidization found in the urine rapidly change from an excess of uric acid, oxalates, and urates to urea, carbon dioxide, and water."

"The quantity of pancreatic juice secreted in twenty-four hours is about 800 grammes, or about thirty-one ounces. It has a very strong alkaline reaction due to the large amount of carbonate of sodium contained in it, which tends to rapidly transform the acid chyme into an alkaline fluid, to convert the proteids into an alkaline albumin, and to place all the food-stuffs in a proper condition to be acted upon by its contained ferments.

"The pancreatic fluid is credited with containing four well-defined ferments and one of questionable existence."

"*First:* the diastatic or amyloptic ferment has for its physiological function the property of converting all forms of starch, both in the cooked or hydrated state and in the raw condition, into glucose ; in intensity of action it is said to stand second only to the diastatic ferment secreted by Brunner's glands. It is by this diastatic action of the pancreatic fluid that the major part of our glucose is formed into starch."

"The *second,* or tryptic ferment, is a powerful erosive agent, attacking alkali albumin in an alkaline medium and converting it first into propeptone, then into albumoses, and finally into a true and diffusible peptone. In this action the proteids do not swell up, gelatinize, and finally become peptones, as in the stomach when acted upon by the pepsine, but the solid proteids are eaten away or eroded, molecule by molecule, until they are finally transformed. In this respect, trypsin differs in action from the pepsin and is very much more powerful, and converts a large quantity of the proteid substance into peptones. Synchronous with this action, numerous by-products are formed, but which, as yet, have become of no practical value."

"The *third,* or milk-curdling ferment, attacks the milk which may have passed through the stomach unaffected by the milk-curdling ferment of the gastric secretion, transforming it into semi-solid and flaky casein to be attacked by the tryptic ferment and converted into a diffusible peptone."

"The *fourth*, fat-splitting or steaptic ferment, is the one by which the various forms of fat are split up into their component parts — fatty acid and glycerine — the former by the further action of the ferment being joined to the sodium bicarbonate of the pancreatic juice, and finally resulting in the formation of a hard soap."

"The *fifth* ferment, emulsin, which is questionable, is probably not a ferment, but the emulsifying action of the pancreatic juice is probably due to the combined action of the strong alkalinity of this fluid and the steapsin on the fats, by which the latter are most perfectly emulsified and placed in a condition in which the epithelial cells can seize upon the minute particles of fat, draw them into the cell protoplasm, and discharge them into the lacteals."

Having demonstrated how this over-feeding disturbs the digestive processes, over-taxes the working capacity of the liver, and exhausts the oxygen capable of being supplied to the animal economy for oxidization purposes, it becomes quite apparent that the functional activity of the liver and pancreatic gland are among the first to be seriously injured, and that their secretions of necessity must be defective both in quantity and quality.

Therefore, only one line of treatment can possibly suggest itself as coming within the limits of physiological or chemical laws.

This is a removal of the old and deteriorated biliary and pancreatic secretions by the use of an active cholagogue, followed by an artificial supply of new bile and pancreatic elements of an improved quality until Nature is capable of producing a good quality by her own inherent powers.

The best method by which the old bile of poor quality can be removed is by the administration of the following "cholagogue tablets, Porter's formulæ.*"

---

* *The Medical News*, Philadelphia, February 28, 1891, p. 45.

"CHOLAGOGINE TABLETS."

℞

Bichloride of Mercury ⎫ of each   1 grain
Arsenious Acid   ⎭
Powdered Ipecac   2 grains
Calomel   15 grains
     M. Divide into thirty tablets, of which one to two may be given every four hours until six to twenty have been taken.

These tablets may be used in different ways. If a quick and positive action is required, one may be administered every three hours until twelve have been taken, or until the bowels commence to move quite freely. If a little slower action is more desirable two tablets may be administered every night for three nights, and then two every second or third night. If a still milder action is needed one or two tablets may be administered every third night.

At the same time that the cholagogue tablets are being taken the following compound Bili-pancreatine pill should also be administered three times a day, preferably about fifteen minutes before the ingestion of food or nourishment, if the most desirable effects are to be attained.

COMPOUND PANCROBILIN PILLS;

OR, COMPOUND BILI-PANCREATINE PILLS.

Porter's Formulæ. *

℞   STRONG.

Inspissated Ox-bile	40 grains
Pancreatic Extract	10 grains
Compound Extract of Colocynth	20 grains
Sulphate of Quinine	20 grains
Extract of Taraxicum	30 grains

     M. Divide into forty capsules, of which one should be taken three times daily before meals.

℞   MILD.

Inspissated Ox-bile	40 grains
Pancreatic Extract	10 grains
Compound Extract of Colocynth	10 grains
Sulphate of Quinine	20 grains
Extract of Taraxicum	30 grains

     M. Divide into forty capsules, of which one should be taken three times daily before meals.

---

* *The Medical News*, Philadelphia, February 28, 1891, p. 45.

℞   COMPOUND AND TONIC.

Inspissated Ox-Bile	40 grains
Pancreatic Extract	10 grains
Compound Extract of Colocynth	20 grains
Sulphate of Quinine	20 grains
Extract of Nux Vomica	6 grains
Extract of Damiana	30 grains

M.   Divide into twenty capsules, of which one to two should be taken three times daily before meals.

Whenever this line of treatment is found to cause the bowels to move too frequently the strong Bili-pancreatine pill should be replaced by the mild or compound pill or the liquid preparation, either of which will still supply the requisite quantity of bile and pancreatic extract and furnishes the necessary aid to the digestive functions without producing hypercatharsis.

The great aim in this method of treatment is to so regulate the actions of the bowels that the patient will have one or two smooth but well-formed evacuations from the bowels each day.

In this connection, it may be well to state that a simple Pancrobilin pill is also recommended containing two grains of inspissated ox-bile and one-half grain of pancreatic extract, and also a liquid preparation of Pancrobilin containing one and one-half grains of pancreatic extract and one grain of bile to the half ounce.

INDEX.

INDEX.

THE FOOD PRODUCTS.

THE NEW AND VALUABLE PREPARATIONS
MANUFACTURED
BY

REED AND CARNRICK.

OR————

CARNRICK'S KUMYSS TABLETS.

A PRODUCT OF PURE, SWEET MILK, PALATABLE, NUTRITIOUS, EASILY DIGESTED, AND WHEN DISSOLVED IN WATER FORMS A DELICIOUS EFFERVESCENT KUMYSS.

(Put up in air-tight bottles, in two sizes, the larger holding sufficient Tablets for seven twelve-ounce bottles, and the smaller sufficient for three twelve-ounce bottles of Kumyss.)

THIS PREPARATION is presented to the Medical Profession in the convenient form of Tablets, and will be found superior in every respect to ordinary *Kumyss, Wine of Milk, Fermented Milk*, or any similar preparation. The following are the points of superiority:

1st. When prepared for use, KUMYSGEN contains every important constituent found in the best *Kumyss* that has been produced in this country or elsewhere. It is made from fresh, sweet milk, guaranteed absolutely free from all contaminations.

2d. It is more palatable and elegant than the ordinary preparations of Kumyss and is always uniform and stable.

3d. In ordinary Kumyss but fifteen per cent. of casein is soluble, while in *Kumyss* prepared from KUMYSGEN fully thirty per cent. of the casein is soluble, and consequently does not tax the digestive organs so much, and is more readily assimilated.

4th. It is less expensive, costing about one dollar and a quarter for each large package of KUMYSGEN, containing sufficient tablets for seven bottles of the same size and strength as the ordinary *Kumyss*, which usually sells for a dollar and seventy-five cents.

5th. There is no loss from breakage of bottles, as KUMYSGEN is ready for use as soon as the Tablets are dissolved.

6th. It is always uniform in acidity, as acid fermentation cannot occur as long as the tablet is kept dry. It is, of course, well known that the ordinary *Kumyss* constantly increases in acidity, and soon becomes so extremely acid that it is unfit for use.

7th. KUMYSGEN being in dry form and secured in air-tight containers, will keep perfectly, and can be safely shipped to any part of the globe, while liquid *Kumyss* cannot be transported any distance unless packed in ice.

8th. KUMYSGEN contains some free and unused ferment, and should a Physician desire an increase of alcohol, lactic acid, and carbonic acid gas it is only necessary to place the bottle of prepared *Kumyss* on its side two to four days at the ordinary house temperature, when the fermentation will continue and develop an increased amount of alcohol and acidity, and will also increase the solubility and digestibility of the casein. This prolonged fermentation, however, is seldom desirable or required.

The use of KUMYSGEN will obviate the constant complaints of patients that *Kumyss* is never twice alike; one day mildly acid, the next day very acid, and the third day very much like vinegar. The acidity, alcohol, and carbonic properties can be perfectly regulated.

By reference to Section I. it will be readily seen that KUMYSGEN is available in the treatment of almost every condition, and that it acts as a highly nutritious and easily digestible food-product.

There is, perhaps, no other dietetic preparation more widely known, more highly esteemed, or more generally indicated in diseased conditions, than the preparation known as KUMYSS. It would be difficult to name a pathological condition in which its use would not be beneficial. Clinical tests gathered from every quarter of the globe attest its special value in all cases of *Gastric* and *Intestinal Indigestion* or *Dyspepsia*, *Pulmonary Consumption*, *Constipation*, *Gastric and Intestinal Catarrh*, *Fevers*, *Anæmia*, *Chlorosis*, *Rickets*, *Scrofula*, *Hepatitis*, *Cystitis*, *Pleuritis*, *Scurvy*, *Vomiting in Pregnancy*, *Bright's Disease*, *Intestinal Ailments of Infants*, *Cholera Infantum*, *for young children*, *and for convalescents from all diseases*.

The casein being finely subdivided, it is especially valuable for all who require an easily digested or a partially digested Food.

KUMYSGEN is a delicious Food-beverage, relished alike by the sick or well.

KUMYSGEN is tonic, stimulant, diuretic, slightly diaphoretic, highly nutritious, easily digested, perfectly palatable, and always *permanent* and *uniform* in strength.

DIRECTIONS.

Put twelve tablets in a clean bottle (conveniently a common beer bottle with the patent rubber stopper) holding about twelve fluid ounces. Any other *strong* bottle of different size may be used, with a good cork well secured. Use one tablet to each ounce of water.

Fill the bottle within an inch of the top with drinking water, either ice-cold or at ordinary temperature, and stopper the bottle tightly at once. Lay the bottle on its side and shake occasionally until the tablets have dissolved, requiring from fifteen to twenty minutes. Drink during effervescence.

Always put on ice or immerse the bottle in cold water and cool before opening the bottle, as at ordinary temperature some of the contents might be lost by too lively effervescence.

Should there be *too much effervescence* to suit the taste and convenience of the patient, the proportion of tablets may be lessened.

Some patients may require a larger proportion of *alcohol, lactic,* and *carbonic acids,* and also find this increased fermentation more agreeable to the palate; for them we recommend a further fermentation, which is accomplished by putting aside the recently prepared Kumyss (laying the bottle on the side) in a room of ordinary temperature (from 70 to 80 degrees) for from two to four days, the former for very hot weather the other for moderate temperatures.

SAMPLE SENT ON REQUEST.

REED & CARNRICK, Manufacturing Chemists,

NEW YORK.

PANCROBILIN

OR,

BILI-PANCREATINE.

IN PILL OR LIQUID FORM.

Each pill contains two grains pure bile and one-half grain pure Extract Pancreas. DOSE: One to three pills before meals.

Each one-half ounce of liquid contains two grains pure Extract Pancreatine and pure bile. DOSE: One to two tablespoonfuls before meals.

PANCROBILIN PILLS, COMPOUND.

* Porter's Formulæ.

STRONG.		MILD.	
Each pill contains:		Each pill contains:	
Inspissated Ox-bile	1 gr.	Inspissated Ox-bile	1 gr.
Pancreatic Extract	¼ gr.	Pancreatic Extract	¼ gr.
Compound Extract of Colocynth	½ gr.	Compound Extract of Colocynth	¼ gr.
Sulphate of Quinine	½ gr.	Sulphate of Quinine	½ gr.
Extract of Taraxicum	¾ gr.	Extract of Taraxicum	¾ gr.
DOSE: One pill three times daily before meals.		DOSE: One pill three times daily before meals.	

COMPOUND AND TONIC.

Each pill contains:

Inspissated Ox-bile	1½ grs.
Pancreatic Extract	¼ gr.
Compound Extract of Colocynth	2-5 gr.
Sulphate of Quinine	¾ gr.
Extract of Nux Vomica	1-6 gr.
Extract of Damiana	¾ gr.

DOSE: One pill three times daily before meals.

The more recent investigations in relation to clinical medicine, and the comparative study of the physiological laws governing the animal economy as applied to the investigation of therapeutic agents, has clearly demonstrated that in bile alone the physician has one of the most powerful remedial agents known to medicine. It has also been quite clearly proven that in all conditions of ill-health or with the introduction of the various toxic agents which are constantly gaining access to the system, that the bile-producing power of the liver is one of the first to become abnormal. The bile secreted becomes defective both in quantity and quality, digestion and assimilation decidedly impaired, the secretory and excretory work of all the glandular organs incompletely performed. The pancreatic secretion, like the bile, becomes decidedly poor in quality.

To meet these defects in the physiological economy,

* The Medical News, Philadelphia, February 28, 1891, p. 45.

which are so universally present in all the diseased conditions of the body and particularly in connection with the intestinal and hepatic derangements commonly called dyspepsia or indigestion, this preparation, known as *Pancro-bilin* or Bili-pancreatine, has been especially designed.

The combination of a considerable quantity of the bile with a smaller quantity of pancreatic extract supplies a fresh stock, and a good quality of the most essential digestive agents needed by the system to thoroughly perform the digestive work. Agents are supplied by this preparation which Nature is not able to construct for herself and without which the best work cannot be accomplished.

On account of the varied conditions which may be associated with this disturbance in bile-producing power and the imperfect pancreatic action, it has been found that no single formula could be made which, upon a scientific basis, would reach all these abnormalities.

Consequently four kinds of pills are manufactured so that these abnormal conditions can be more effectually reached, and it is confidently believed that by a judicious selection from these four preparations the *plain,* the *compound, mild, strong,* and *tonic* Pancrobilin pills or liquid preparation almost every physiological abnormality requiring this combination can be most effectually treated.

When the physiology of digestion and assimilation is perfectly understood, there is scarcely a condition which will not be favorably influenced by some one of these three preparations ; sometimes one will be needed and sometimes another while treating the same patient.

It is hardly necessary to state that *Pancrobilin* is rapidly increasing in popularity among the Medical Profession. It is being used in many cases with the happiest results where treatment for stomach indigestion has previously failed. In such cases the trouble has arisen in the intestines and liver which are now becoming recognized as the local points of origin for the vast majority of the so-called Gastric Dyspepsias.

One of the most essential features in connection with the therapeutic use of *Pancrobilin* is to first remove all the old and deteriorated bile from the system and then supply some new bile of a good quality.

REED & CARNRICK, Manufacturing Chemists,

NEW YORK.

Cholagogine Tablets.

EACH TABLET CONTAINS :

Bichloride of Mercury	1-30 grain
Arsenious Acid	1-30 grain
Powdered Ipecac	1-15 grain
Calomel	½ grain

DOSE : One to two, every four hours, until six to twenty have been taken.

THESE Tablets have been especially designed to remove completely the old and deteriorated biliary and pancreatic secretions from the system, and at the same time to stimulate the liver and pancreatic gland to a more perfect action. The removal of the defective biliary and pancreatic secretions is effected by the action of the bichloride of mercury, arsenious acid, and ipecac, all three of which tend to stimulate a very active and abundant discharge of biliary and pancreatic fluid from the glandular epithelium of the liver and pancreatic gland, while at the same time the small and frequently repeated dose of calomel contained in the Tablet causes a very copious discharge of this hyper-secreted bile which is contained in the bile ducts, and ultimately from the system, in connection with the stools. The more general utility of these Tablets will be found in the text of parts *one* and *two*.

Put up in bottles holding 100 and 1,000 tablets.

Diureticine Tablets.

THIS COMBINATION HAS BEEN LARGELY USED AS A NON-IRRITATING DIURETIC.

EACH TABLET CONTAINS :

Sulphate of Strychnine	1-60 grain
Hydrochlorate of Caffeine	1 grain
Powdered Digitalis	½ grain

DOSE : One to three, every three or four hours.

THIS Tablet is especially applicable to those cases of renal inactivity which are due to a weak-acting heart, a low tension in the blood-vessels, and general venous engorgement.

In the Bright's of alcoholic subjects it is very efficacious. It often has to be given for two or three days to get the best evidence of its activity in exciting the renal glands to increased action.

Put up in bottles holding 100 and 1,000 tablets.

REED & CARNRICK, MANUFACTURING CHEMISTS,
NEW YORK.

INNERVATINE TABLETS.

EACH TABLET CONTAINS:

Sulphate of Strychnine	1-60 grain
Hydrochlorate of Caffeine	1 grain
Extract of Damiana	1½ grains
Extract of Taraxicum	½ grain

DOSE: From one to two, every four or five hours.

THESE TABLETS ARE DESIGNED ESPECIALLY AS ADJUNCTS TO THE CHOLAGOGINE TABLETS AND PANCROBILIN PILLS IN THE TREATMENT OF CHRONIC DIGESTIVE DERANGEMENTS 🐚

ALSO in connection with all cases in which the nerve force is greatly depressed. Long continued malnutrition of the system develops a depressed and neurasthenic condition of system.

In these cases the lack of vital force is at such a low ebb that, even though the food chosen be perfect and the most available digestive agents are used, the patients steadily fail. The administration of this Innervatine Tablet furnishes to the system what it cannot generate for itself, and is often — together with a good diet and the Pancrobilin pills — the factor needed to effect a complete recovery.

Put up in bottles holding 100 and 1,000 tablets.

REED & CARNRICK, MANUFACTURING CHEMISTS, NEW YORK.

Cardiene Tablets.

THIS Tablet, which is composed of three of the most commonly known and generally used alkaloids, is most highly recommended as an agent by which the nutritive vitality of the heart can be augmented. It also slows and steadies the action of the heart and increases the relative force of the heart's action. All these alkaloids, being of nitrogenous composition, can be utilized by the system for constructive purposes, or at least to retard largely tissue waste. By a steady use of this Tablet, a compensatory hypertrophy can be developed, which is often equal to the mechanical deficit in the heart, and thus the patient is made practically well.

Put up in bottles holding 100 and 1,000 tables.

REED & CARNRICK, MANUFACTURING CHEMISTS,

NEW YORK.

ANALGESINE TABLETS.

EACH TABLET CONTAINS:

Hydrochlorate of Cocaine	½ grain
Phenacetine	1½ grains
Exalgine	1½ grains
Salicylic Acid	1½ grains

DOSE: From one to three, every four hours, until pain is relieved.

THIS Tablet, composed as it is of some of our most powerful pain-killing, medicinal agents, has been found to reach many cases which heretofore required the use of opium. It will be very efficacious in treating all forms of neuralgias, and particularly that of gastric origin. It also relieves occipito-frontal, facial, intercostal, and general neuralgias. In many instances it will relieve the pain of a sciatica. The neuralgic pains of ovarian origin during menstruation and at other times are often completely relieved by a few of these tablets.

Put up in bottles holding 100 and 1,000 tablets.

REED & CARNRICK, MANUFACTURING CHEMISTS,
NEW YORK.

HYPNOTINE TABLETS.

EACH TABLET CONTAINS:

Abstract of Hyoscyamus	1-20 grain
Chloralamide	1 grain
Sulphonal	1 grain
Extract of Damiana	1½ grains

DOSE: From one to three, every three or four hours.

This hypnotic Tablet is designed as an adjunct to the general treatment of digestive difficulties, and to relieve the troublesome insomnia so common with hepatic derangements.

The great advantage of this preparation is that when taken regularly it produces natural sleep, without depressing the nervous system, and without a tendency to develop a habit for some sleep-producing agent. It helps to improve the nutritive condition of the nervous system at large, and soon its use can be entirely suspended.

Put up in bottles holding 100 and 1,000 tablets.

CARNRICK'S FOOD.

The analysis of this preparation will show that its chemical constituents are almost *identical* with an average sample of *human milk.*

Formula for Carnrick's Food.—CARNRICK'S FOOD, as now prepared, is composed of the solid constituents of cow's milk, of wheat (the starch being converted into soluble starch and dextrine) and sufficient milk sugar to make the total percentage equal to that of human milk.

There is no other food for infants or children in the market that so closely resembles human milk in the proportion of its constituents excepting Lacto-Preparata.

It is put up in hermetically sealed cans, and being sterilized will keep indefinitely, and reach the hands of the nurse or mother free from every contamination.

REED & CARNRICK, MANUFACTURING CHEMISTS,
NEW YORK.

LACTO-PREPARATA.

FOR INFANTS
DURING THE NURSING PERIOD.

The Only Practically All Milk Food Ever Prepared; Corresponds Almost Perfectly with Healthy Human Milk in Composition, Digestibility, and Taste.

Made only from pure milk obtained from cows fed with perfect food and pure water, all of which are under the supervision of one of the best milk experts in this country; consequently no danger can arise from diseased milk.

Lacto-Preparata is put up now ONLY in air-tight cans, and will keep perfectly.

It is treated during the process of manufacture according to the directions of Prof. ATTFIELD for STERILIZING MILK.

Our new factory, where the milk is reduced to a powder in vacuo, is located at Bainbridge, Chenango Co., N.Y. The elevation of the surrounding country is the highest in New York State except the Catskill and Adirondack region, and consequently the milk is pure and free from zymotic germs.

Lacto-Preparata is the most perfect Infant Food that can be prepared, at least for the period that a child should be nourished upon milk exclusively. It is far superior to any method of preparing cow's milk in the household, and, what is more important, the milk is free from all danger of germ disease, fermentation, and other abnormal conditions incident to handling milk, especially in cities.

REED & CARNRICK, MANUFACTURING CHEMISTS,

NEW YORK.

Lacto-Cereal Food.

MILK, FRUIT, AND CEREALS.

A PERFECT INVALID FOOD.

LACTO-CEREAL FOOD

Is composed of powdered milk (sterilized and partly digested), starch, — dextrinated wheat, malted barley, desiccated bananas, cacao-butter, and manna. A small percentage of parched corn is added, which gives it an appetizing and elegant flavor. Unlike other powdered foods, it contains the ferment that digests starch.

LACTO-CEREAL FOOD

Is a highly-nutritious, perfectly palatable, and very easily digested food. It is designed for **invalids, dyspeptics, convalescents, consumptives, the aged,** and all persons whose stomachs are weak and who require nourishment in its simplest and least irritating form, as in cases of **fevers, gastric ulcers,** and **cancer of the stomach.**

LACTO-CEREAL FOOD

Will be found of great value if used as part of the daily diet of children and of aged people, when the digestion is feeble and their strength and vitality need recuperation.

PUT UP IN POUND AND HALF POUND CANS.

REED & CARNRICK, MANUFACTURING CHEMISTS,
NEW YORK.

CORRIGENT PILL.

FORMULA.

Hypophosphites Iron, Manganese, Lime, Soda, and Potash, each ⅛ grain; Hypophosphite Quinine, ½ grain; Hypophosphite Strychnine, 1-200 grain; Purified Sulphur, ⅔ grain. DOSE: one to three pills after meals.

Indicated especially in Neuralgia, General Debility, Malaria, Neurasthenia, and Phthisis.

When the body is deficient in strength, or "run down" from any cause, this restorative pill will be found invaluable in imparting tone to the system, increasing the appetite, improving the digestion, enhancing the functions of assimilation and blood-making, and removing malarial and other taints from the blood.

Put up in bottles containing 100 and 1,000 pills.

CORDIAL ANALEPTINE.

FOR RHEUMATIC AND GOUTY DIATHESIS.

FORMULA.

Each tablespoonful contains Sodium Sulphide, 1-10 grain; Salicylate of Lithia, 2 grains; Acetate of Potash, 5 grains; Ex't Black Cohosh, 3 minims; Ex't Cascara Sagrada, 3 grains. DOSE: dessert to tablespoonful as often as indicated.

The Salicylates have especial control over the manifestations of acute rheumatism, as Black Cohosh and Cascara have over the subacute and more chronic forms. Lithia removes the excess of uric acid and purifies the blood, and acetate of potash assists by stimulating the kidneys and increasing the flow of urine. The sulphur increases the secretion of bile and stimulates the excretory action of the liver.

Put up in fourteen-ounce bottles.

REED & CARNRICK, MANUFACTURING CHEMISTS,
NEW YORK.

Sulpho-Calcine.

Sulphur-Tartrate Tablets.

Each Tablet contains three grains Sulphur, one grain Cream of Tartar, and one grain Powdered Milk Sugar.

These Tablets will be found of great value in Biliousness, Liver Affections, Skin Diseases, and Chronic Muscular Rheumatism.

As Sulphur is an important constituent of the nails and hair, the irregular growth and formation of the former and the loss of the latter are undoubtedly due to the loss of this element in the system.

Dose, one to two Tablets at bedtime, or one night and morning. They should be taken for a length of time to obtain full effect.

Put up in bottles holding 100 and 1,000 tablets.

REED & CARNRICK, MANUFACTURING CHEMISTS, NEW YORK.

VELVET-SKIN SOAP.

THE PERFECTION OF TOILET SOAPS.

It is composed of the following ingredients :

Best Lagos Palm Oil, Pure Olive Oil, Bay Berry Tallow, Oat Meal Flour (finely bolted), Witch Hazel Extract, Slippery Elm Bark (mucilaginous portion) and Borax.

Velvet-Skin Soap is made from *vegetable* oils. There is, therefore, no necessity of an excess of alkali, and the soap is made neutral and mild, with perfect freedom from any liability to rancidity.

VELVET-SKIN POWDER.

THE PERFECTION OF ALL BABY AND TOILET POWDERS.

It is composed of

Venetian Talc, Orris Root, Lycopodium, and Boracic Acid.

Velvet-Skin Powder does not cake like chalk or starch powders, and is harmless.

It will be found of great service in excoriated or sore nipples, and all irritated surfaces.

Velvet-Skin Powder is an excellent remedy for Scalds, Burns, Hives, Rashes, Chapped Hands, Prickly Heat, and Dandruff.

REED & CARNRICK, Manufacturing Chemists, NEW YORK.

www.ingramcontent.com/pod-product-compliance
Lightning Source LLC
Chambersburg PA
CBHW021825190326
41518CB00007B/745